Deckneri

ACS Symposium Series

M. Joan Comstock, *Series Editor*

FOREWORD

The ACS SYMPOSIUM SERIES was founded in 1974 to provide a medium for publishing symposia quickly in book form. The format of the Series parallels that of the continuing ADVANCES IN CHEMISTRY SERIES except that in order to save time the papers are not typeset but are reproduced as they are submitted by the authors in camera-ready form. Papers are reviewed under the supervision of the Editors with the assistance of the Series Advisory Board and are selected to maintain the integrity of the symposia; however, verbatim reproductions of previously published papers are not accepted. Both reviews and reports of research are acceptable since symposia may embrace both types of presentation.

Introduction
to Microlithography

Introduction to Microlithography

Theory, Materials, and Processing

L. F. Thompson, EDITOR
Bell Laboratories

C. G. Willson, EDITOR
IBM Research Laboratory

M. J. Bowden, EDITOR
Bell Laboratories

Based on a workshop

sponsored by the ACS Division

of Organic Coatings

and Plastics Chemistry

at the 185th Meeting of the

American Chemical Society,

Seattle, Washington,

March 20–25, 1983

ACS SYMPOSIUM SERIES **219**

AMERICAN CHEMICAL SOCIETY
WASHINGTON, D. C.   1983

Library of Congress Cataloging in Publication Data

Introduction to microlithography.
  (ACS symposium series, ISSN 0097-6156; 219)

  Includes bibliographies and index.
  Contents: An introduction to lithography/L. F.
Thompson—The lithographic process—the physics/
L. F. Thompson and M. J. Bowden—Organic resist
materials—theory and chemistry/C. Grant Willson—
[etc.]

  1. Photolithography—Congresses. 2. Photoresists—
Congresses.

  I. Thompson, Larry F., 1944–   . II. Willson, C.
Grant, 1939–   . III. Bowden, Murrae J. S.,
1943–   . IV. American Chemical Society. Division of
Organic Coatings and Plastics Chemistry. V. American
Chemical Society. National Meeting (185th: 1983:
Seattle, Wash.) VI. Series.

TR940.I57   1983   686.2′315   83–5968
ISBN 0–8412–0775–5

CONTENTS

PREFACE

THE REMARKABLE PROGRESS in the miniaturization of electronic devices over the past three decades promises to continue for many years. At the heart of this revolution is the silicon integrated circuit (IC) whose complexity and performance continue to increase, paving the way for continued innovation in solid state devices. The improvements in IC performance stem primarily from an ability to internally interconnect more and more active components (transistors, diodes, etc.) on a single "chip" of silicon. In 1965, 250 devices could be interconnected in the "monolithic integrated circuit" and in 1983 over 1,000,000 devices are routinely connected in a single device. This improvement has come about largely through a decrease in the size of the circuit elements. The size (area) of the chips has not changed significantly over the years, whereas the size of each element has decreased from >20 μm in 1963 to <2.0 μm in 1983.

The three-dimensional circuit elements are fabricated by a series of processes collectively known as "lithography". The pattern is first generated in a polymeric film on a device "wafer" and this pattern is then transferred via etching, into the underlying thin film. The purpose of this book is to review the theory, materials, and processes that are used in the lithographic process. This book is intended to be a tutorial and not a comprehensive review. Each chapter contains many references to which the reader can refer for more detail on any specific aspect of microlithography.

The authors are indebted to the Bell Laboratories Text Processing group, especially T. I. Howard, J. Alder, and M. McCann. Special thanks also go to J. H. Bruning for his assistance with Chapter 2.

L. F. THOMPSON
Bell Telephone Laboratories
Murray Hill, NJ 07974

C. G. WILLSON
IBM Research Laboratories
San Jose, CA 95193

M. J. BOWDEN
Bell Telephone Laboratories
Murray Hill, NJ 07974

February 18, 1983

An Introduction to Lithography

L. F. THOMPSON

Bell Laboratories, Murray Hill, NJ 07974

0097-6156/83/0219-0001$06.00/0
© 1983 American Chemical Society

1.1 Historical Perspective

Soon after the invention of the point contact transistor over three decades ago, several important technological trends were rapidly established. First, a technology known as planar junction technology (*1*) was developed; the development of this technology permitted the construction of p-n junctions in a semiconductor material without the use of bulky, three-dimensional contact devices. Within a few years, silicon had become the dominant material for semiconductor fabrication, and photolithography was established as the technique of choice in fabricating planar, silicon devices. Numerous volumes have been written about the chemistry and physics of semiconductor devices and of materials for semiconductor manufacturing (*2-4*). Much less information has been compiled concerning the processing steps involved in semiconductor device fabrication, especially in the area of lithography.

Another important innovation in semiconductor devices was the invention of the monolithic integrated circuit. This technology provides a means for the *internal* interconnection of many solid-state and passive devices into a working unit that performs complex, electronic functions and allows the "density" of active devices to be increased many thousands of times over that achievable with discrete or single transistor devices. The past two decades have seen an astronomical growth in the microelectronics industry and associated technologies. The application of microelectronics impacts technologies such as communications, computer science, medicine, energy, and more recently, home entertainment. This revolution is going to continue for the foreseeable future and will require continued advances in silicon technology. The realization of this new generation of devices must involve the development of new processing, fabrication, control, and manufacturing technologies.

The conventional way of achieving these goals has been to make devices smaller, this results in more active-circuits per unit area (*5*). Figure 1 illustrates this trend, and it is the conviction of the author that this will continue for at least two more decades. It is interesting to note that not only are we decreasing the size of the elements in an integrated circuit while increasing the number of active devices, but that the average cost per chip is also remaining essentially constant -- that is the cost per function is decreasing dramatically. This trend has required and will continue to require a greater understanding of device physics, device design, and lithography. Microcircuit fabrication requires the selective diffusion of tiny amounts of impurities into specific regions of the semiconductor substrate to produce the desired electrical characteristics of the circuit. These regions are defined by lithographic processes in which the desired pattern is first defined in a resist layer (usually a polymeric film which is spin-coated onto the substrate) and

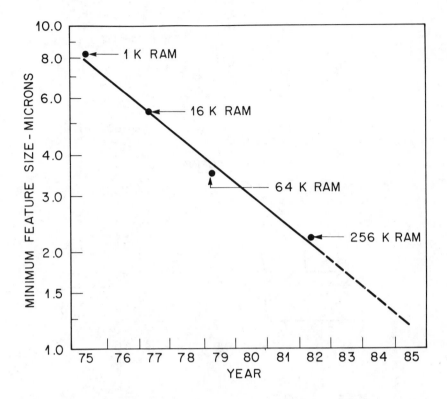

Figure 1.   Minimum feature size on a MOS random access memory device as a function of the year the devices were first commercially available.

subsequently transferred, via techniques such as etching, ion implantation and/or diffusion, to the underlying substrate. This process is shown schematically in Figure 2. The purposes of this book are: (a) to review the lithographic strategies currently being used and under development, and (b) to emphasize in particular, the importance of resists and resist processing in achieving each individual lithographic technology goal.

1.1.a Solid State Devices.   Many *designs* and *strategies* for device design and fabrication have intervened between the point contact transistor and the planar technology to which we alluded earlier, including grown junction transistors, alloyed junction transistors, and many variations of the planar technology.   An understanding of the definition of integrated circuits, together with the details of the physics involved in their operation, is an

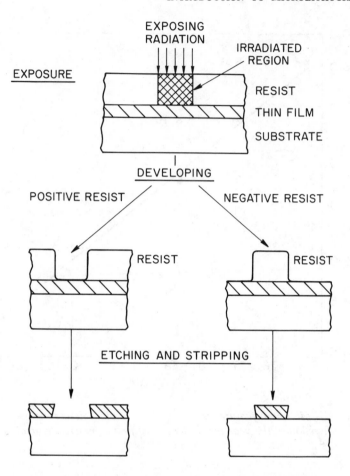

Figure 2. The schematic diagram showing the formation of a polymeric relief image using lithography. The resist pattern is used to subsequently modify the underlining substrate.

interesting story in itself, but beyond the scope of this book whose objective is limited to the lithographic process involved in circuit manufacturing (6). A single transistor is essentially a junction between two types of semiconducting materials (denoted p and n) and metal connections to the outside world. A large-scale integrated circuit contains tens of thousands of these individual transistor elements that are interconnected in complex ways by conductors such as aluminum or highly doped polycrystalline silicon. A simple transistor element of an integrated circuit is depicted in Figure 3.

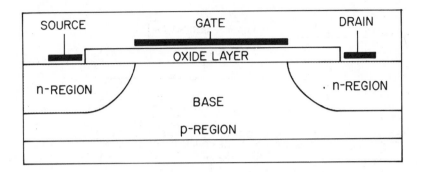

Figure 3. Schematic representation of a single transistor similar to those used in a MOS memory device.

Silicon is the dominant material for microelectronic circuits, primarily because of the ease with which it oxidizes to form insulating barriers for the subsequent implanting of tiny amounts of dopants into selected regions to achieve the requisite electrical properties. The silicon dioxide insulator and other dielectric films that are commonly encountered such as silicon nitride films are patterned by a process known as photolithography. Photolithography is probably *the* key process in microelectronic fabrication technology, because it is repeated 5 — 12 times before the three-dimensional circuit geometries necessary for a completed metal oxide semiconductor (MOS) or bipolar device are achieved. Figure 4 is an outline of the manufacturing sequence of a large-scale integrated circuit and illustrates the importance of understanding the lithographic technology used to delineate the patterns of thin-film dielectrics and conductors. The structure of an integrated circuit is complex both in the topography of its surface and in its internal composition. Each element of such a device has an intricate three-dimensional architecture that must be reproduced exactly in every circuit. The structure is made up of many layers, each of which is a detailed pattern. Some of the layers lie within the silicon wafer and others are stacked on the top. The manufacturing process consists in forming this sequence of layers precisely in accordance with the plan of the circuit designer.

A concept for a new circuit is transformed into a circuit design by engineers who have a knowledge of both circuit electronics and processing. Circuit designers who conceive of the new product work at specifying the functional characteristics of the device and select the processing steps that will be required to manufacture it. The actual design of the device begins

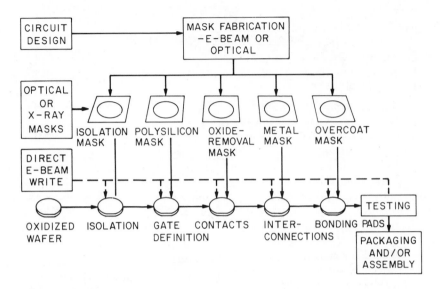

Figure 4.   Outline   of   Manufacturing   Sequences   for   a   typical   solid   state   device.

with an estimation of the size and approximate location of every circuit element. This preliminary design work is usually done with the aid of computers. Upon completion of the circuit design layout, a computer memory contains a list of the exact position of every element in the circuit. From data in the computer memory a set of plates, called photomasks, is prepared. Each mask contains the pattern for a single layer of the circuit. Since the circuits are so small, many can be fabricated side by side simultaneously on a single wafer of silicon. Thus each photomask, typically a glass plate about five inches on a side, has a single pattern repeated many times over its surface.

A complete set of correct masks is the culmination of the design phase of the development of the microelectronic circuit. The plates are delivered to the wafer-fabrication facility, where they will be used to produce the desired sequence of patterns in a physical structure.

As the complexity of integrated circuits increases and the dimensions of each circuit element decrease, greater demands are placed on the lithographic process, particularly with respect to resolution. Indeed, we are rapidly approaching the limits of photolithographic techniques for reproducing circuit geometries. We should also point out that in addition to the exposure technology, the resist itself is also an important factor limiting resolution and must be considered as we continue the trend of decreasing element size and increasing integration complexity. In subsequent chapters we will examine the various alternatives to photolithography, as well as the importance of processing and resists in this industry.

1.1.b Lithography, the Art.   Lithography, often referred to as photoengraving, is the process of transferring a pattern into a reactive polymer film (termed a resist) which will subsequently be used to replicate that pattern into an underlying thin film or conductor. Photolithography, which uses ultraviolet radiation (360 — 410 nm), is the current "workhorse" technology in integrated circuit fabrication. It is interesting to note that the technology currently used has been in existence since the early days of planar transistor development (7). Many of the processing steps, which were worked out empirically with the use of phenomenological models and parametric studies, fall more into the realm of art rather than science.

The understanding of the physics and chemistry of resist materials and attendant processing has been largely neglected in the three decades of semiconductor device innovations. As the minimum feature size in a device approaches 1 μm and new lithographic technologies develop rapidly, it is of paramount importance that we transform the important area of lithographic processing from an art to a science. Thus an understanding of the fundamental physics and chemistry of resist exposure, together with the compatibility of semiconductor processing steps with polymer films, is of crucial importance in extending semiconductor device geometries into the submicron region.

1.2 Lithographic Strategies

Integrated circuit fabrication requires a method for forming accurate and precise patterns on silicon substrates. These patterns delineate the area for subsequent doping and/or internal interconnection. As mentioned earlier, this is currently done by photolithography and is shown schematically in Figure 5. The photolithographic process consists of (a) producing a mask carrying the requisite integrated circuit pattern information for a given level and (b) subsequently transferring that pattern, using some optical technique into a photoactive polymer known as a resist. Typically, ultraviolet light in the wavelength region 350 — 430 nm is used with a variety of exposure techniques to be discussed later. Diffraction considerations limit the size of the individual elements to about 1 μm, and new lithographic strategies will be required to achieve the dimensions and overlay accuracies required for future generations of devices. Several strategies are currently being pursued

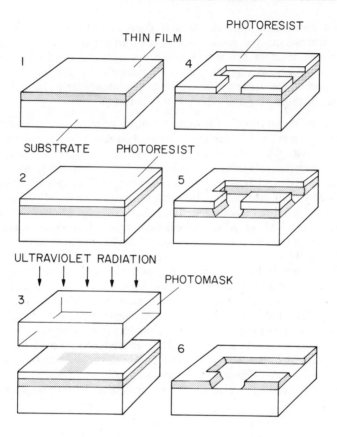

Figure 5.   A typical photolithography process

to achieve this goal. These include: short wavelength photolithography (8-10), electron beam lithography (11-17), x-ray lithography (18-20) and ion beam lithography (27-30). The final choice as to which strategy will be adopted will depend not only on the capabilities in terms of performance, resolution and registration accuracy, but also on the cost effectiveness involved in manufacturing integrated circuits. At the present time, the lithographic process contributes only a small fraction (less than 10%) of the cost of a finished device. However, if a lithographic technology that requires large capital investments in exposure equipment with low throughput is chosen, the lithographic process could become a dominant fac-

tor in the cost of submicron devices. Each of the potential lithographic strategies has its own inherent limitations which must be evaluated; these
include resolution, registration accuracy, throughput, capital investment in
exposure hardware, and others.

1.2.a Photolithography. The most common photolithographic technique is
contact printing (*8*). This involves holding a mask just off the wafer surface
and visually aligning the mask to the previous pattern on the wafer. This
process is known as registration or alignment. After alignment is achieved,
the mask is pressed into hard contact with the resist coated wafer which is
then exposed through the mask with a flood beam of ultraviolet light. A
modification of this technique is to maintain a space between the mask and
the wafer. This technique, known as soft contact printing or proximity
printing (*21-23*), minimizes mask/wafer damage caused by contact, but at
the expense of resolution. A recent development in photolithography is the
technique known as projection alignment, in which the mask image is projected onto the wafer through a reflective optical system (*8,24,26*). This
technique improves mask lifetime and decreases defect densities because
intimate contact between the wafer and mask is not required. Because
wafers are increasing in size every few years, a continuing problem is the
task of designing optics that can form an accurate image over larger and
larger areas. Another photolithographic technique which has been
developed involves exposing a smaller area (on the order of 1 sq. cm.) and
stepping this pattern over a large — diameter wafer. This technique (known
as step-and-repeat photolithography) improves resolution and alignment
accuracy.

The minute size of the features that can be formed using photolithography is ultimately limited by the *wavelength* of the exposing radiation.
Current photolithographic techniques can routinely reproduce geometries a
few microns across, and it is likely that dimensions as small as 1 μm can be
produced using step-and-repeat photolithography. The resolution achieved
by these photolithographic techniques can be improved by using shorter
wavelength UV light.

1.2.b X-ray Lithography. In many laboratories x-ray lithography has been
investigated as a possible replacement for photolithography. X-rays with a
wavelength of a few angstroms used in an exposure format similar to contact or proximity photolithography have been used to fabricate features as
small as 0.02 μm (*27,30*). Since x-ray lithography allows exposure of the
entire wafer, the process is potentially inexpensive. However, many problems still exist, including the fabrication of stable x-ray masks, alignment,
and the availability of suitable, commercially available resist systems.

1.2.c Electron Beam Lithography. Electron beam lithography is a mature technology, which developed from scanning electron microscope developed in the early 1960's (*31-36*). Systems for electron beam exposure are very similar to scanning electron microscopes although the complexity of data transfer and computer-control systems needed for electron beam writing machines results in a cost of several million dollars per machine. Electron beam writing is capable of submicron pattern definition and is used currently by many companies to manufacture specialty, high-resolution devices. It is also used for the fabrication of photomasks that are used with one-to-one and step-and-repeat projection photolithographic systems. The principal drawback of electron beam lithography is low throughput coupled with high capital cost. However, progress in this field continues at a rapid pace, and no doubt less expensive, higher throughput machines will be developed in the next two decades. Two major advantages of electron beam lithography are (a) its ability to register accurately over small areas of a wafer and (b) lower defect densities. The latter advantage results from the lack of a need for intermediate masks.

1.2.d Ion Beam Lithography. Ion beam lithography has been investigated recently by several workers (*18-20*). Compared with electron beam lithography, it offers the advantage of improved resolution (ions are scattered much less than electrons during passage through the resist) as well as lower resist sensitivities. The hardware for this technique is similar to that of an electron beam exposure system, the major difference being the source. Before this technique can be competitive with electron beam lithography, however, advances must be made in the areas of high brightness sources and high speed deflection systems.

In subsequent chapters we will briefly review the limitations and advantages of each of these technologies, and we will relate the advantages and disadvantages specifically to the resist and subsequent processing that must be employed to utilize each of these technologies.

1.3 Resist Materials and Processes

Two families of photoactive and/or radiation-sensitive materials have been described in the literature: organic and inorganic. The inorganic materials, described several years ago by Yoshikawa and co-worker (*37,38*), are based on silver-sensitized chalcogenide glasses and are sensitive to almost all forms of actinic radiation. These materials offer considerable promise for photolithography and have been described in detail by Tai et al. (*39*). Organic materials are generally polymeric in nature, with properties tailored for a specific lithographic technology. This book will deal exclusively with organic-based radiation-sensitive materials that are used in lithographic processes.

Photoresist and photoengraving processes have been described and used in industry for well over 100 years. In 1826, W. H. F. Talbott received a British patent (No. 565) on a photoetching process used to engrave copper. The resist material was gelatin, sensitized with a bichromate salt; ferric chloride was the etchant. In 1852, J. N. Niepce discovered that certain types of asphalt were sensitive to ultraviolet radiation, and he applied these materials successfully to the etching of pewter and other decorative metals. In the ensuing years, many natural products (including protein, shellac, starches, and carbohydrates) were sensitized with dichromate and bichromate salts. These materials were all negative-acting resists *(7,40)*.

In 1958, P. J. Griess discovered that aromatic orthoquinone diazids would undergo reactions when exposed to ultraviolet light. Almost 100 years elapsed before these materials were used in positive acting resist systems. In the early 1920s, Kalle and Company, AG in Germany, and its American subsidiary Azoplate Corporation developed the first positive-acting photoresist based on novolac resins and quinone diazid photoactive dissolution inhibiters *(41)*. Originally this work was aimed at the fabrication of lithographic plates, and only later were these systems applied to microcircuit manufacturing. It is interesting to note that the basic chemistry and formulations have undergone little change over the ensuing years and that these resist systems still constitute the "workhorse" materials of the semiconductor industry.

The first modern day negative photoresists were developed by the Eastman Kodak Company which utilized cyclized rubbers and cinnamic acid derivatives as photosensitive crosslinking agents *(42)*. The first commercially important photoresist based on this chemistry was known as KPR, which was of a cinnamate ester of polyvinyl alcohol. It was introduced by Kodak in 1954.

Before the invention of the planar transistor, many photoresist processes were developed for the manufacture of circuit boards. Experience gained in this area was rapidly transferred to silicon processing, and much of the early work in integrated circuit lithography can be traced directly to circuit board manufacturing.

The basic concepts employed in early photolithography, both in materials and processing, have been extrapolated to modern lithographic technologies including x-ray and electron beam. In the remainder of this book we will discuss in detail the fundamental principles of chemistry and physics as they apply to the design and use of resist materials.

Literature Cited

1. Grove, A. S. "Physics and Technology of Semiconductor Devices"; John Wiley and Sons: New York, 1967.
2. Forester, T. "The Microelectronics Revolution", The MIT Press: Cambridge Massachusetts, 1980.
3. *Scientific American*, September 1977, *237*, (3). (Special issue on Microelectronics development.)

4. Bylander, E. G. "Materials For Semiconductor Functions", Hayden Book Co.: New York, New York, 1971.
5. Moore, G. E. *Proceedings of the IEEE,* 1976, *64* (6), 307.
6. Young, E. S. "Fundamentals of Semiconductor Devices"; McGraw-Hill, New York, New York, 1978.
7. Deforest, W. S. "Photoresist Materials and Processes"; McGraw-Hill, New York, New York, 1975.
8. Doane, D. A., *Solid State Technol.,* 1980, *23* (8), 101.
9. King, M. C.; Muraski, E. S. *Proc. Soc. Photo-opt. Instrum. Eng.,* 1979, *174,* 70.
10. King, M. C. *IEEE Trans. Electron Devices,* 1979, *ED-26,* 711.
11. Herriott, D. R.; Brewer, G. R. "Electron-Beam Lithography Machines" in *Electron-Beam Technology in Microelectronic Circuit Fabrication*; Academic Press, New York, New York, 1981; pp 141-216.
12. Varnell, G. R.; D. F. Hebley,; Robbins, R. R.; Carpenter, C.; Malone, J. J. *J. Vac. Sci. Technol.* 1980, *16,* 1787.
13. Herriott, D. R.; Collier, R. J.; Alles, D. S.; Stafford, J. W. *IEEE Trans. Electron Devices* 1975, *ED-22,* 385.
14. Alles, D. S.; Ashley, F. R.; Johnson, A. M.; Townsend, R. L. *J. Vac. Sci. Technol.* 1975, *12,* 1252.
15. Yourke, H. S.; Weber, E. V. *Tech. Digest, IEDM* 1976, Washington, pp 431-436.
16. Weber, E. V.; Moore, R. D. *J. Vac. Technol.* 1979, *16.,* 1780.
17. Pfeiffer, H. C. *J. Vac. Sci. Technol.* 1978, *15,* 887.
18. Broers, A. N. *Physics Today* 1979, *11,* 38.
19. Seliger, R. L.; Kubena, R. L.; Ulney, R. D.; Ward, J. N.; Wang, V. *J. Vac. Sci. Technol.* 1979, *16* (6), 1610.
20. Hall, T. M.; Wagner, A.; Thompson, L. F. *J. Vac. Sci. Technol.* 1979, *16,* 1889.
21. Kaneko, T.; Umegaki, T.; Kawakomi, Y. *Proc. Kodak Interface 80* 1980, *25.*
22. Caprari, F.; Doane, D. A. *ECS Abs.* 1980, *80* (2), 858.
23. Massetti, D. O.; Hockey, M. A.; McFarland, D. L. *Proc. Soc. Photo-opt. Instrum. ENG,* 1980, *32,* 221.
24. Markle, D. A.; *Solid-State* Technol, 1974, *17* (6), 50.
25. Cathbert, J. D. *Solid State Technol.* 1077, *20* (8), 59.
26. Bruning, J. J. *J. Vac. Sci. Technol.* 1980, *17,* 1147.
27. Spears, D. L.; Smith, H. I. *Electron Lett.* 1972, *8,* 102.
28. Feder, R.,; Spiller, E.; Topalion, J. *J. Vac. Sci. Technol.* 1975, *12,* 1332.

29.   Fay, B.; Trotel, J.; Petroff, Y.; Pinchaux, R.; Thiry, P. *Appl. Phys. Lett.* 1976, *29*, 370.

30.   Maldonado, J. R.; Coquin, G. A.; Maydan, D.; Somekh, S. *J. Vac. Sci. Technol.*, 1975, *12*, 1329.

31.   Pfeiffer, H. C. *Record Symp. Electron, Ion, Laser Beam Technol.* (1974), 11th, Boulder, Colorado, May 1974, San Francisco Press, San Francisco, California.

32.   Broers, A. N. *Proc. Symp. Electron, Ion Beam Sci. Technol.* 1972, pp 3-25.

33.   Broers, A. N. *J. Vac. Sci. Technol*, 1973, *10*, p 979.

34.   Alles, D. S.; Ashley, F. R.; Johnson, A. M.; Townsend, R. L. *J. Vac. Sci. Technol.*, 1975, *12*, p 1252.

35.   Herriott, D. R.; Collier, R. J.; Alles, D. S.; Stafford, J. W. *IEEE Trans. Electron Devices,* 1975, *ED-22*, p 385.

36.   Varnell, G. L.; Spier, D. F.; Rodger, A. C.; Holland, R. D. *Proc. Int. Conf. Electron Ion Beam Sci. Technol. 6*th, San Francisco, California, p 97.

37.   Yoshikawa, A.; Ochi, O.; Nagai, H.; Mizushima, Y., *Appl. Phys. Letters*, 1977, *31* (3), p 161.

38.   Yoshikawa, A.; Ochi, O.; Nagai, H.; Mizushima, Y. *Appl. Phys. Lett.,* 1976, *29* (10), p 677.

39.   Tai, K. L.; Sinclair, W. R.; Vadimsky, R. C.; Moran, J. M., *J. Vac. Sci. Technol*, 1977, *16*, p 1977.

40.   Jorgensen, G. W.; Bruno, M. H., "The Sensitivity of Bichromated Coatings", 1954, *Bull.* 218, Lithographic Tech. Foundation, New York, New York.

41.   Kosar, J., "Light Sensitive Systems"; John Wiley & Sons, New York, New York, 1965; p 194.

42.   Minsk, L. M. US Pat. 2,725,372 (1955).

43.   Schlaback, T. D.; Rider, D. K., "Printed and Intergrated Circuitry"; McGraw-Hill Co., New York, New York, 1963; p 85.

RECEIVED February 4, 1983

The Lithographic Process: The Physics

L. F. THOMPSON and M. J. BOWDEN

Bell Laboratories, Murray Hill, NJ 07974

0097-6156/83/0219-0015$16.80/0
© 1983 American Chemical Society

Circuit fabrication requires the selective diffusion of tiny amounts of impurities into specific regions of a semiconductor substrate to produce the desired electrical characteristics of the circuit. These regions are defined by lithographic processes which consist of two steps: (1) delineation of the desired pattern in a resist layer (usually a polymeric film which is spin coated onto the substrate) and (2) transfer of that pattern via process such as etching into the underlying substrate. The faithful replication of the pattern originally described by the device designer is limited by the physical and chemical processes involved in both stages of pattern transfer. In this chapter we will confine our attention to the first stage of the lithographic process, viz., pattern delineation in the resist (the second stage is discussed in Chapter 5) and in particular, to the fundamental physical limitations of each of the exposure technologies as they relate to sensitivity and resolution.

The primary definition of the circuit pattern in the resist is itself a two-stage process consisting of the formation of a latent image by the exposure tool, followed by development of that image to produce a three-dimensional relief structure. Thus a consideration of factors that determine resolution in the primary pattern definition step requires an understanding of the physics and chemistry of both the exposure and the development process in order to distinguish between those resolution limitations associated with latent image formation and those due to resist processing. In the following sections we will consider the physical factors that limit resolution associated with exposure hardware and resist exposure. Resist processing is discussed in Chapter 4.

There is a variety of exposure technologies, each with a fundamental resolution limit resulting from radiation interactions with both the hardware and the resist. From a practical standpoint, factors such as alignment tolerance and resist swelling usually degrade resolution to a point where minimum feature sizes are larger than theoretically expected. Nevertheless, examining the fundamental resolution limits of the various techniques can be instructive.

2.1 Photolithography

There are many methods of producing patterns in photoresists utilizing ultraviolet light (200-400 nm). The earliest of these was called *contact* or *proximity printing* and simply entailed flood exposing a wafer coated with a photosensitive polymer with ultraviolet light through a mask as shown in Figure 1 (*1,2*). The mask contains clear and opaque features that define the circuit pattern. These techniques (which are also collectively known as *shadow printing*) have been the mainstay of device fabrication for many years.

2.1.a Contact Printing. The primary resolution limitation of shadow printing is diffraction of light at the edge of an opaque feature on the mask as the light passes through an adjacent clear area (*3-6*). Figure 2 illustrates a

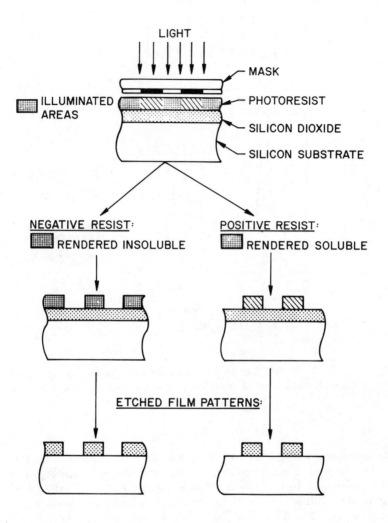

Figure 1. Schematic of contact or proximity printing using positive and negative resists.

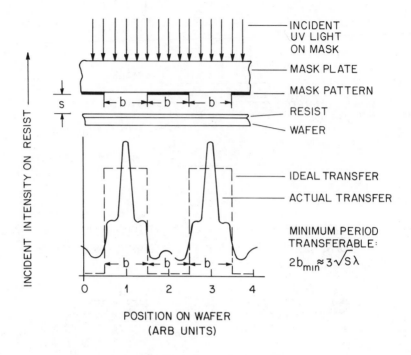

Figure 2. Light distribution profiles on a photoresist surface after light has passed through a mask containing an equal line and space grating.

typical intensity distribution of light incident on a photoresist surface after passing through a mask containing a periodic grating consisting of opaque and transparent lines or spaces of equal width, b. It is obvious from this figure that diffraction causes the image of a perfectly delineated edge to become blurred or diffused at the resist surface. Just how the light is "bent" (i.e., diffracted) as it passes an opaque edge will be discussed in section 2.1.d.

The theoretical resolution capability of shadow printing (using a conventional photoresist) with a mask consisting of equal lines and spaces of width b is given by (2)

$$2b_{min} = 3\sqrt{\lambda(s + \frac{1}{2}d)} \tag{1}$$

where 2*b* is the grating period (1/2*b* is the fundamental spatial frequency *ν*), s the gapwidth maintained between the mask and the photoresist surface, λ the wavelength of the exposing radiation and d the photoresist thickness. For hard contact printing, *s* is equal to 0, and thus from Equation 1, the maximum resolution for 400 nm, wavelength light and a 1 *μ*m thick resist film will be slightly less than 1 *μ*m. Within diffraction limits, contact printing (so called because the mask and wafer are in intimate contact) can transfer a mask pattern into a photoresist with almost 100% accuracy and is the form of optical imaging that provides the highest resolution. Other photolithographic techniques can approach but not exceed its resolution capabilities.

As noted earlier, the maximum resolution capability is seldom achieved because of other considerations such as resist swelling, difficulty in alignment, and debris between mask and wafer. The debris results from contact between mask and wafer that causes damage to both the mask and the resist surface and introduces undesirable defects (7,8). These defects are reproduced in all subsequent exposures, thus resulting in decreased device yields. Particles present between the mask and wafer also prevent intimate contact and degrade resolution across the wafer. The lack of flatness in either mask or wafer will create in-plane distortions on contact leading to misalignment errors and thereby degrading further the useful resolution capability of this technique. Registration requires the mask to be moved relative to the wafer. This motion introduces debris or particles that further complicate defect problems.

2.1.b Proximity Printing. The defect problem associated with contact printing may be reduced by introducing a gap (*s* in Equation 1) between the mask and the wafer. However, increasing the gap degrades resolution by expanding the penumbral region caused by diffraction. For any gap the minimum transferable period is given by Equation 2.

$$2b_{min} \approx 3\sqrt{s\lambda} \tag{2}$$

As an example, the maximum resolution for a 10 - *μ*m gap, using 400 - nm exposing radiation will be $\approx$ 3 *μ*m.

Proximity printing requires a small and very constant spacing which can be achieved only with extremely flat wafers and masks (9). In the past several years, techniques associated with wafer and mask fabrication and polishing have improved markedly so that the average flatness today is on the order of a few microns over the entire wafer and/or mask. This has not only improved overlay accuracy but also has extended the useful resolution capabilities of contact and proximity printing by allowing smaller spacings. A gap of 10 *μ*m is the smallest gap that can be used in practice in order to assure that the mask and wafer never come in contact with each other. We may note from Equations 1 and 2 that the resolution for shadow printing is proportional to the square root of the wavelength. Thus it should be possible to improve resolution by decreasing the wavelength of the exposing radiation. This will be discussed further in section 2.1.e.

In spite of the limitations of contact and proximity printing, these technologies are widely used, primarily because of their relatively low cost and ease and simplicity of operation. Several companies use these techniques to produce a variety of solid state devices including VLSI memory.

2.1.c Projection Printing. The problems and limitations of contact and proximity printing have led to the development of projection printing techniques in which lens elements are used to focus the mask image onto a wafer substrate which is separated from the mask by many centimeters. Because of lens imperfections and diffraction considerations, projection techniques generally have lower resolution capability than that provided by shadow printing. However, the net improvement in device yield and performance as a result of lower defect densities and improved registration has led to the increasing importance of these techniques in VLSI production.

Several types of projection printing strategies have been developed and are shown schematically in Figure 3. One of the most successful approaches utilizes a 1:1 wafer scan (Figure 3a) exemplified by the Perkin-Elmer Micralign scanning projection printer which uses *reflective* spherical mirror surfaces to project images of the mask onto the wafer substrate (*10-12*) (Figures 4, 5 and 6). The illuminating radiation is generated from a standard mercury lamp and passes through a condenser and slit a few millimeters in width. This slit or arc of radiation is imaged by primary and secondary mirrors to cover the entire width of the mask, as shown in Figure 4. The wafer and mask are scanned through this arc of radiation by means of a continuous scanning mechanism (Figure 5). This scanning approach is designed to minimize distortions and aberrations of the optical system by keeping the imaging illumination always within the zone of good optical correction. The projection optics used in Micralign systems form an image on the wafer that is a mirror image of the mask, the same inversion that occurs in contact printing. Continued improvements in optical and mechanical design have resulted in a technology that is capable of producing 1.0 to 1.5 μm features with an overlay accuracy of better than $\pm$ 0.40 μm using 350-400 nm light.

The resolution of a projection system is given by the relationship

$$w = \frac{k\lambda}{NA} \tag{3}$$

where w is the minimum feature size, k an empirically determined constant that depends on photoresist parameters and processes, λ the exposing radiation wavelength, and NA the numerical aperture of the optical system (see Section 2.1.d.). These machines have a throughput greater than 40 wafers

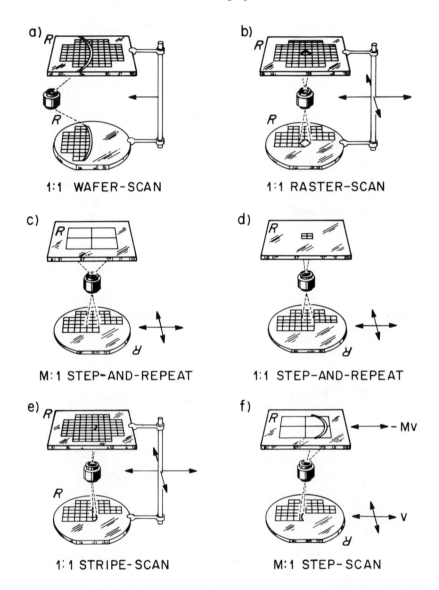

a) 1:1 WAFER-SCAN

b) 1:1 RASTER-SCAN

c) M:1 STEP-AND-REPEAT

d) 1:1 STEP-AND-REPEAT

e) 1:1 STRIPE-SCAN

f) M:1 STEP-SCAN

Figure 3. Schematic of present and potential future optical lithography systems: (a) Perkin Elmer Micralign (10), (b) Bell Labs printer (11), (c) reduction step-and-repeat (Censor, Electromask, GCA, Optimetrix, Philips), (d) 1X step-and-repeat (Ultratech), (e) 1X stripe scan, and (f) reduction step-scan. "R" indicates object and image orientations. Lenses are indicated only schematically. (Reproduced with permission from Ref. 30)

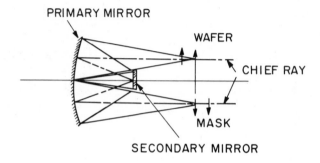

Figure 4. Ray diagram for the Perkin-Elmer Micralign projection system

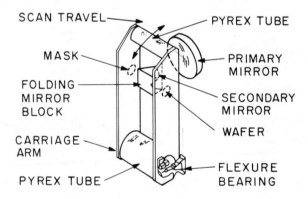

Figure 5. The scanning stage subsystem in the Perkin-Elmer Micralign projection system.

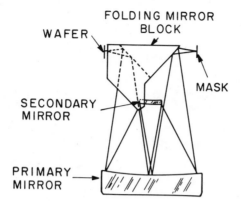

Figure 6. Schematic of the Perkin-Elmer Micralign optical system.

per hour and represent the "workhorse" lithographic tool in many semiconductor factories.

A second type of exposure system uses refractive optics similar to that used in conventional photographic cameras. Refractive lenses capable of producing a one-to-one image over an entire three to four inch diameter silicon wafer would need to be extremely complex requiring many elements, and would be prohibitively expensive (*13*). On the contrary, refractive optical systems employing 10:1 and 5:1 reduction of a reticle (Figure 7) are practically and economically feasible. Because of their better resolution and registration capability, these systems are achieving widespread use in the semiconductor industry for fabricating devices with features less than 1.5 μm. This approach requires a precise mechanical stage which can step the image over the entire wafer surface with consequent reduction in the number of wafers that can be printed per unit time. In spite of the reduced throughput, step-and-repeat projection printing is currently one of the fastest growing technologies in photolithography (*14-20*).

2.1.d. Physical Limitations of Optics. The resolution associated with the optics of projection systems is diffraction limited, i.e., the optical elements are sufficiently perfect that their imaging characteristics are dominated only by diffraction effects. In order to understand the resolution limitations of projection printers, one must understand the concepts of coherency, image formation, diffraction and numerical aperture (*3,9,11,12,14*).

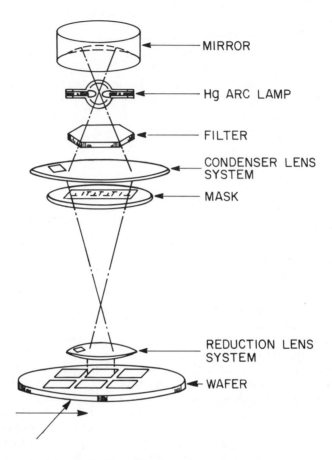

Figure 7. Schematic of a reduction step-and-repeat system.

2.1.d.1 coherency. The UV light source of a projection printer has several important parameters that affect performance: 1) intensity (brightness), 2) wavelength, 3) bandwidth, and 4) coherency (*21-23*). All sources have a finite size and produce a finite amount of energy per unit area. If we view the source, depicted in Figure 8, from point P at a distance ℓ, we see a large area emitting radiation whose spectral output is characteristic of the source. The wavelength is determined by the temperature and the type of material used as the emitter. For example, tungsten emits photons in the infrared

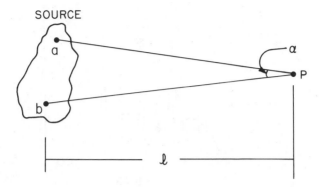

Figure 8. An incoherent source of optical radiation of finite size.

frequency range at low temperature to near UV frequencies at higher temperatures. Discharge arcs are used for most lithographic sources with the wavelength being dependent on the type of gas and pressure in the arc confinement. Typical gases are mercury and xenon which emit many wavelengths with a spectrum characteristic of the particular source. The spectrum is usually composed of many sharp lines and is not of uniform intensity at every wavelength. A typical spectral distribution for a high pressure mercury arc source is shown in Figure 9. Wavelengths of a specific frequency and bandwidth can be obtained from a source by using filters.

Coherent is defined as "the state or quality of being together," and when applied to photons falls into two categories, *temporal* and *spatial* (*3, 21, 24*). Temporal coherency refers to photons that are related to each other in *time*, i.e., they are emitted simultaneously and are related in time at all equivalent points on the same wavefront. The source depicted in Figure 8 can be subdivided into an infinite number of *point* sources (such as *a* and *b* in Figure 8) each emitting *independently* of the other. Since there is no way each point source can "know" how any other point is emitting, it is impossible for a finite source to be temporally coherent. As it turns out, temporal coherency is not important from the standpoint of lithographic performance for most sources of practical interest. Far more important is spatial coherency which refers to the *phase* relationship of one photon to another. A spatially coherent source is one whose photons are in phase at every point along any wave front with the wave being described by a uniform mathematical function at any time or any position. Figure 10 illustrates a perfectly coherent source radiating a series of spherical wave fronts.

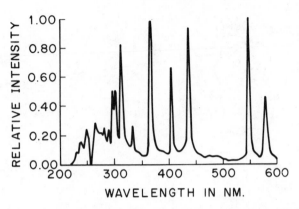

Figure 9. Typical high pressure Mercury-arc spectrum.

By definition, a wavefront is the locus of points, all of which are in the same phase. Thus when an observer at point P "looks" at any two photons along the paths P_1 and P_2 at points 1 and 2 on a single wave front, those two photons will have the same phase.

A laser is spatially coherent as is a conventional source that is infinitely small. Referring to Figure 8, this may be achieved by moving the observation point P to infinity, at which point α, the angle subtended at P, approaches zero as does the area of emission. We should point out, however, that brightness for a finite source is defined as power per unit area per unit solid angle. Therefore, achieving coherency in this manner reduces the intensity to zero and would require infinite exposure time. Fortunately we do not need perfect coherency, a point that will be treated in more detail later.

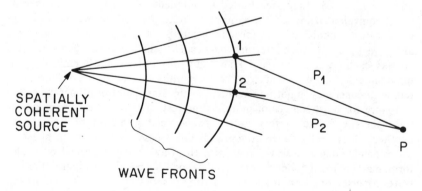

Figure 10. Schematic illustrating a perfectly coherent source.

2.1.d.2 image formation.   The lens is the most important part of a projection system.   It is a device constructed of transparent material whose function is to collect the light emitted from an object (such as a mask) and focus or image this light at the wafer plane (*25*).   Figure 11 illustrates a simple lens that images an object (O) onto the wafer plane (I)

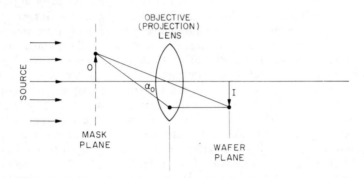

Figure 11. Image formation by a refractive lens system showing the object O in the mask or object plane imaged to I in the wafer or image plane.

where the image is seen to occur at the point where rays of light emanating from a given point are made to converge by the lens.   The lens in an optical projection printer is extremely complex involving many elements.   Nevertheless, the simple lens shown in Figure 11 and succeeding figures will serve just as well for illustrative purposes.   The optical quality of the image is evaluated in terms of resolution (*minimum size* of object that can be imaged over the entire field of the lens), contrast and size of the useful image field. The latter is related to the total number of points that can be imaged side by side over a field.   These factors determine the lithographic performance of a lens (*14*) and are dependent on lens design parameters.

From the standpoint of image projection, it is important to recognize that a lens can reconstruct an image only from light that has been *diffracted* and that diffraction in turn limits the maximum resolution which can be attained.   Thus an understanding of the phenomenon of diffraction is important in order to understand even qualitatively the optics associated with image projection.

2.1.d.3 Diffraction. According to *geometrical* optics, if an opaque object is placed between a point (coherent) light source and a screen as shown in Figure 12a, the edges of the object will cast a sharp shadow on the screen. No light will reach the screen at points within the geometrical shadow, while outside the shadow the screen will be uniformly illuminated. In reality, the shadow cast by the edge is diffuse, consisting of alternate bright and dark bands that extend into the geometrical shadow (see Figure 12b). This apparent bending of light around the edge is referred to as *diffraction*, and the resulting intensity distribution is called a *diffraction pattern.*

Two edges very close together constitute a slit from which very distinct diffraction patterns are produced when illuminated with monochromatic light. The particular intensity distribution observed depends upon whether we are viewing the diffraction pattern under conditions where the source and/or screen on which the pattern is displayed, are located a short distance from the slit (Fresnel diffraction) or at an infinitely large distance (Fraunhofer diffraction). We will immediately recognize that the diffraction limiting resolution in shadow printing falls into the Fresnel class since the mask (slit) and wafer (screen) are separated by only a very short distance. Of the two classes, Fraunhofer diffraction is much more simple to treat mathematically because as we shall see later, the diffracted wave is plane making possible a fairly simple method of summing up analytically the disturbances reaching any point. We will assume Fraunhofer diffraction conditions in the following discussion.

Experimentally it is found that when the width of the slit is narrower than the wavelength, the radiation spreads out (i.e., is diffracted) as it passes through the slit with the intensity being nearly the same at all points on a semicircle with its center at the slit. If the width of the slit is comparable with the wavelength, the beam no longer spreads uniformly in all directions but becomes more concentrated towards the plane through the source and midline of the slit. If the slit is wider than a wavelength, new phenomena appear, the central high intensity beam becoming subdivided by a series of bands of relatively low intensity. Figure 13 shows the Fraunhofer diffraction pattern corresponding to the latter case for two different slit widths and shows how the beam begins to spread out as the slit width is narrowed.

The distribution of light intensity in Figure 13 can be computed by application of Huygens principle which allows us to calculate the shape of a propagating wavefront provided the wavefront at an earlier instant is known. According to this principle, every point of a wavefront may be considered as a source of secondary waves (often called a *wavelet*) which spread out in all directions, i.e., all points on a wavefront are point sources for the production of spherical secondary wavelets. The new wave front $\sum^1$ is then found by constructing a surface tangent to all the secondary wavelets as shown in

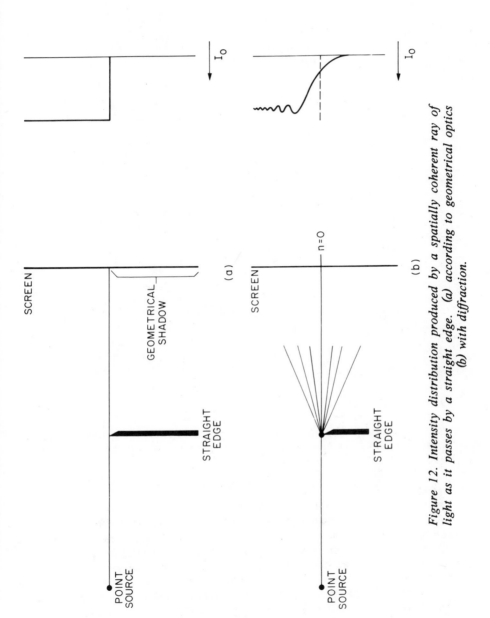

Figure 12. Intensity distribution produced by a spatially coherent ray of light as it passes by a straight edge. (a) according to geometrical optics (b) with diffraction.

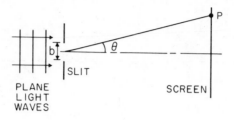

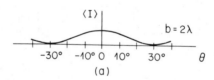

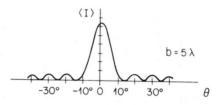

(a)

Figure 13. Fraunhofer diffraction pattern of a single slit illuminated with coherent monochromatic light; the intensity distribution is shown for two different slit widths.

Figure 14a. Provided the wavefront is not interrupted, it moves out uniformly in a radial direction in accordance with the accepted notion of the rectilinear propagation of light. Suppose we now have a series of *plane* wavefronts incident on a slit with each successive wavefront being constructed from the envelope of secondary wavelets from the preceeding wavefront as shown in Figure 14b. If the width of the slit is small relative to the wavelength of light, then the portion of the wavefront at the slit may be considered as a *single* Huygens source radiating in all directions to produce

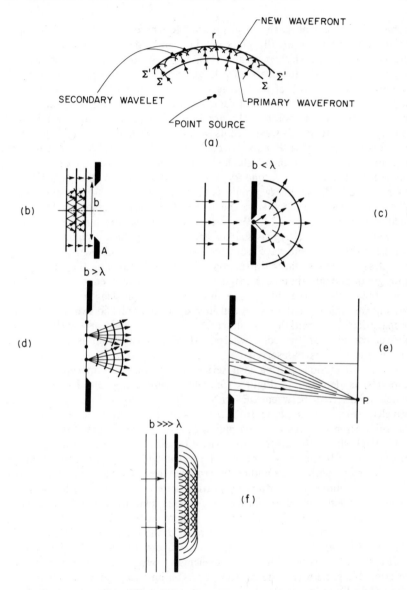

Figure 14. (a) *Schematic of Huygens Principle showing construction of a new wave front* Σ^1 *from the preceeding wave front* Σ; (b) *plane wave front incident on a slit of width b;* (c) *diffraction for case where* $b < \lambda$; (d) $b > \lambda$ *showing subdivision of the slit into a distribution of Huygens sources;* (e) *superposition of waves at P;* (f) *diffraction effects for the case where* $b >>> \lambda$.

a spherical wavefront as shown in Figure 14c where, as indicated before, the intensity is the same at all points on the semicircle whose center is at the slit. When the slit width is greater than a wavelength, we must consider the wavefront at the slit to consist of a large number of Huygens sources distributed across the width of the slit as shown in Figure 14d. Each is of equal brightness, each is in phase, and each generates a wave of a equal amplitude which is detected at point P (Figure 14e). Remember, that when P is at infinity, the diffracted wave fronts are parallel and the rays shown in Figure 14e arriving at P may be considered to be parallel to each other. The intensity at P is then calculated from the superposition of all waves emanating from these secondary sources according to the principles of interference. The intensity profile is then calculated by similar summation of disturbances at all points in the plane of the screen resulting in profiles such as shown in Figure 13. When the slit width is very large compared to the wavelength, the Huygens sources combine to produce a plane wave front parallel to the incident wave front (corresponding to rectilinear propagation) except at the edges where diffraction effects still occur, i.e., the slit width is now great enough that each edge acts as an isolated edge (Figure 14f). Since the width of the slit is large relative to the wavelength, the wavefront passing through the slit may be considered to be infinite in extent and hence propagation is essentially rectilinear. It is only where the wavefront is obstructed, viz., at the edge where part of the wavefront is cut off, that diffraction effects are observed.

A diffraction grating is a series of slits that are spaced at regular intervals as depicted in Figure 15a. The *collimating* lens in Figure 15a which is inserted between the slit and the grating ensures that all the wavelets start out in phase from *all* slits in the grating. The wavefronts emerging from the lens are then planes at right angles to the axis of the system. Each slit in the grating gives rise to a diffraction pattern in which the intensity distribution is a function of the slit width. The diffracted beams then interfere with one another to produce the final pattern. The second lens placed immediately after the grating serves to bring parallel rays to a focus. The lens therefore forms on a screen placed at the focal plane a reduced image of the Fraunhofer diffraction pattern that would otherwise appear on a screen at infinity. Figure 15b shows the diffraction pattern obtained for two, three and four slits respectively for the case where the slit width is much narrower than the wavelength. We immediately note that at certain angles the waves are in phase and combine to produce intensity maxima, while for other angles the phases are such that vector addition of the disturbances results in intensity minima. The major peaks in the intensity distributions in Figure 15b are called *principal maxima* and occur at the same angles for all three gratings. Note how the width of the principal maxima decreases as the number of slits increases. The principal maxima occur at angles θ_N given by

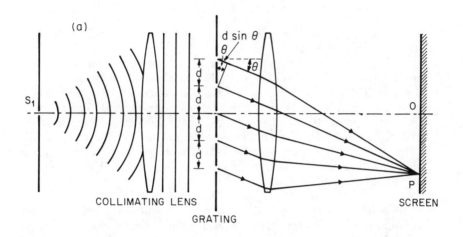

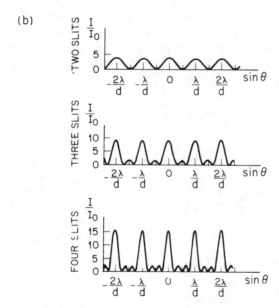

Figure 15. (a) Principle of a grating. (b) Intensity patterns produced by diffraction gratings with two, three and four slits of width <λ with each having the same separation between adjacent slits.

$$\sin \theta_N = \frac{N\lambda}{d} \quad N=0,\pm 1,..... \tag{4}$$

where d is the distance between the slits and λ the wavelength. N is an integer known as the *order*.

It is readily apparent that a mask consisting of equal lines and spaces constitutes a diffraction grating and when uniformly illuminated will produce a diffraction pattern similar to that just discussed with an intensity profile depending on the grating period (ν), the wavelength (λ) and the position of the screen. If the latter is placed at the focal plane of the lens, the diffracted orders will be focussed to form the Fraunhofer diffraction pattern of the grating. These various orders will come together in the image plane where they interfere to produce an image of the mask. This will be discussed in more detail later.

2.1.d.4 Numerical Aperture. Once light passes through a mask and has been diffracted, it is next imaged onto a wafer using an objective lens as depicted in Figure 16. A lens is usually designed to produce an image of magnification M which in turn determines the spatial position of the object and image relative to the lens. Thus the objective lens depicted in Figure 16 will be unable to collect light from angles $> 2\alpha_o$ where α_o is the maximum cone angle of rays reaching the lens from the object point (mask). For a given magnification, α_o is defined by the numerical aperture NA_o of the lens via the expression

$$NA_o = n \sin \alpha_o \tag{5}$$

where n is the refractive index in image space and is usually equal to unity.

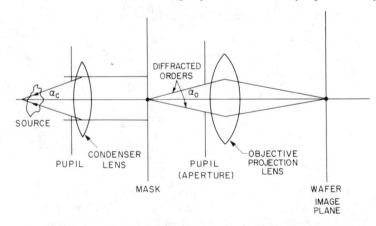

Figure 16. A refractive lens imaging system using partially coherent light, condenser lens and objective lens.

Numerical aperture is thus a quantitative measure of the "acceptance" angle of a lens. Likewise the numerical aperture of the condenser lens in Figure 16 is defined as

$$NA_c = n \sin \alpha_c \qquad (6)$$

where $2\alpha_c$ is the maximum cone angle of rays subtended at the lens on the object side. By combining the concept of numerical aperture with that of diffraction, it is clear from Figure 16 that numerical aperture — and hence the distance from the object (mask) to the lens — determines the number of diffracted orders that are collected and imaged. It is important to reiterate that pattern information is contained only in the diffracted light and not in the zero order beam. Images are formed by the intersection of rays of light which emanate from a point (see Figure 11). The undiffracted or zero order beam from an edge or grating constitutes in effect only a single ray, and at the very least we need a second ray emanating from the edge that can be made to intersect the zero order ray in order to reconstruct an image of the edge. This second ray is provided by the diffracted beam (see Figure 16) which is made to converge with the zero order beam at the image plane via the projection lens thereby forming an image of the edge. A perfect reconstruction of the edge would require the collection of *all* diffracted orders. It is apparent then that the larger the NA of the projection lens, the greater is the amount of diffracted information that can be collected and subsequently imaged. As we will show later, this has important implications to resolution since lenses with higher $NA's$ also have higher resolution.

We have now discussed qualitatively several concepts which are important in developing an understanding of image projection in optical lithography. These are summarized below:

1. Spatially coherent light is light that has a specific phase relationship between each photon on wave fronts emitted from the source.

2. Diffraction refers to the apparent deviation of light from rectilinear propagation as it passes an obstacle such as an opaque edge. We use the term "apparent" deviation because in fact the light is not deviated but arises as a natural consequence of the way in which light is propagated (Huygens Principle). Diffraction effects occur whenever there is any limitation of the width of a beam of light.

3. Spatial information about an object is contained *ONLY* in the diffracted light.

4. A lens images the diffracted light onto a wafer with the amount of information collected from the mask by the lens being related to the numerical aperture of the lens with higher NA lenses having higher resolution capability.

We will now take these qualitative ideas and treat them quantitatively.

The concept of modulation transfer function (MTF) is used to describe and predict the resolution capabilities of projection aligners. MTF is a measure of the accuracy of image transfer with respect to sharpness or contrast. We will choose as our model for discussing MTF, a mask pattern consisting of a grid of periodic opaque lines and transparent spaces of equal widths (b) as shown in Figure 17. The diffraction of light at the edge of an opaque feature results in the projected pattern or image in the photoresist exhibiting gradual (rather than sharp) transitions from light to dark. The edges of the projected feature therefore appear blurred rather than sharp. As the size of b approaches dimensions on the order of 1 μm, the imaged "opaque lines" have considerable light in the middle of the feature. Modulation is defined by $M = \dfrac{I_{max} - I_{min}}{I_{max} + I_{min}}$ where the I_{min} is the light intensity in the center of the dark line, and I_{max} is the light intensity in the center of the bright area. We then define the MTF of an exposure system as the

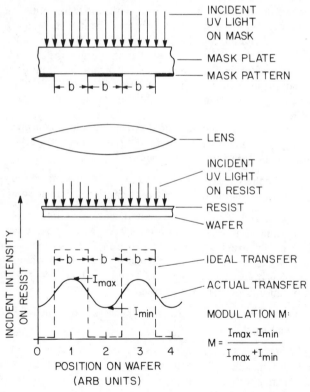

Figure 17. Schematic representation of image transfer efficiency for a 1:1 projection printer. (Reproduced with permission from Ref. 1)

ratio of the modulation in the image plane to that in the object plane. The MTF is dependent on the spatial frequency of the object to be imaged, the numerical aperture (NA) of the optical system (defined by Equation 4), conditions of illumination, and wavelength of the exposing radiation (*1-3,27-29*).

The modulation vs. frequency curve for a given NA and wavelength is shown in Figure 18. Strictly speaking calculations of MTF as a function of spatial frequency express the modulation in the image plane for a *sinusoidal* object whereas in practice a mask represents a binary or square wave object. A sinusoidal object gives rise to a first-order diffraction pattern only whereas a binary object such as a grating produces an infinite number of orders. Perfect image reconstruction requires the collection of *all* diffracted orders which, from the mathematical standpoint, is relatively simple for a sinusoidal object since there is only one order to collect. For large mask features ($> 3 \ \mu$m) sufficient diffracted orders are collected that the mathematical treatment based on a sinusoidal object represents a close approximation. As seen in Figure 18, modulation is dependent on the degree of spatial coherency. Just why this should be so will be examined later.

Two types of illumination are used in projection printers, Köhler illumination and critical illumination shown in Figures 19a and b respectively. With Köhler illumination,

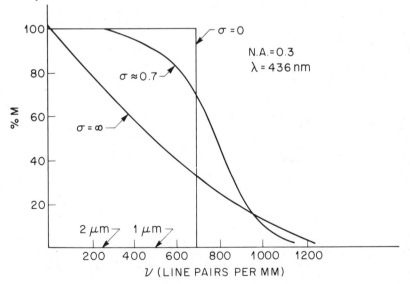

Figure 18. Modulation of an image as a function of the spatial frequency, ν for three coherency factors, σ = 0, 0.7, and ∞.

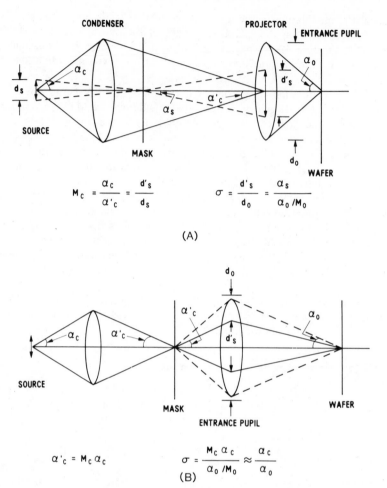

Figure 19. Two common types of illumination systems: (a) Köhler, (b) critical.

the exposure source is imaged through a condenser *in the entrance pupil* of the projection lens (see Figure 19a). The image of the source is the effective size of the source $d_s{}'$, and the degree of coherency, σ, is given by

$$\sigma = \frac{d_s{}'}{d_o} = \frac{\alpha_s}{\alpha_o M_o} \tag{7}$$

where M_o is the magnification of the objective lens, d_s' is the size of the source *image* and d_o is the size of the entrance pupil. The magnification of the condenser lens, M_c, is given by,

$$M_c = \frac{\alpha_c}{\alpha_c'} = \frac{d_s'}{d_s} \tag{8}$$

where the terms are defined in Figure 19a.

In critical illumination, the source is imaged at the mask plane (Figure 19B), and the coherency is given by

$$\sigma = \frac{\alpha_c / M_c}{\alpha_o M_o} \tag{9}$$

where the terms are defined in Figure 19b. If $M_c = M_o = 1$ (as in the Perkin-Elmer Macralign) then

$$\sigma = \frac{d_s'}{d_o} = \frac{\alpha_c}{\alpha_o} \tag{10}$$

The entrance pupil is nothing more than the area of a lens that accepts radiation and defines the relative useful portion of the lens. Since a coherent source is a point source (zero diameter), the image in the entrance pupil will also have zero diameter and thus $\sigma = 0$. An incoherent source is rigorously a source of infinite extent ($\sigma = \infty$) and is only of academic interest since all light collected from any real source is always imaged within the entrance pupil, so that σ is always $\leqslant 1$. In practice, partially coherent sources are used.

Figure 20 depicts the diffraction of coherent and incoherent light by a grating pattern of frequency $\nu = (2b)^{-1}$. The solid lines in Figure 20 show a coherent ray, incident normally on the grating pattern. The dashed line represents an incoherent ray incident on the grating at an angle ϕ and with its first-order diffracted ray emerging at the same angle θ as the coherent ray. We should point out here that in order for the direction of the two first-order diffraction maxima to be the same, it is necessary that the grating frequency be different for both cases under consideration since the angle of a ray diffracted from a grating is dependent *only* on the wavelength and not the incident angle. Clearly, if the grating frequency were the same, the direction of the first diffracted order for the incoherent case would be different from that of the coherent ray.

We will remember that the undiffracted, or d-c component of light emerging from the grating contains no information by itself about the spatial frequency or image position. This information is contained in the diffracted light of orders $N \geqslant 1$. The direction of the diffracted rays for the coherent case is given by the standard grating formula (cf. Equation 4)

$$n_0(2b) \sin \theta = N\lambda \qquad (11)$$

where n_0 is the refractive index in image space (assumed to be 1), b the grating spacing (making the frequency $\nu = (2b)^{-1}$), θ the angle of the ray of order N emerging from the grating, and λ the wavelength of the exposing radiation. Thus the grating frequency ν that results in a first-order diffracted peak at angle θ_1, is obtained as

$$\nu = n_0 \frac{\sin \theta_1}{\lambda} \qquad (12)$$

Since it is not necessary to collect diffracted orders > 1 in order to recreate the image, we will confine our discussion to the first-order diffracted ray. If the light diffracted in direction θ_1 is to reach the image plane through the optical system, θ_1 must be $\leq \alpha_0$ (see Figure 16) where α_0 is defined by $NA_0 = n_0 \sin \alpha_0$ where NA_0 is the numerical aperture of the projection lens. Light scattered at angles > α_0 will not pass through the entrance pupil of the projection lens since it is outside the maximum cone angle of acceptance of the lens. From these considerations, the highest grating fre-

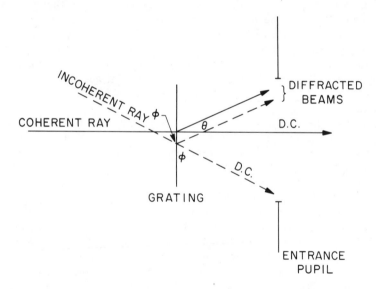

Figure 20. Diffraction of coherent and incoherent light from a grating pattern with a spatial frequency, $\nu = (2b)^{-1}$.

quency that can be imaged by an optical system with coherent illumination is given by

$$\nu_{max} = n_0 \sin \alpha_o / \lambda = NA / \lambda \qquad (13)$$

Further, for a sinusoidal grating, collection of the first-order diffracted peak is sufficient to recreate a perfect image so that the modulation remains at 100% up to the cut-off frequency (cf. Figure 18). For incoherent illumination, incident at an angle ϕ, the direction of the intensity maxima is given by a more general equation

$$n_0 (2b)(\sin\phi + \sin\theta) = N\lambda \qquad (14)$$

where the left side represents the path difference for light passing through adjacent slits (for light incident normal to the grating, $\phi = 0$, $\sin\phi = 0$ and Equation 14 reduces to Equation 11). In order for both the first-order diffracted beam and the d-c beam to reach the image plane, *both* ϕ and θ must be $< \alpha_0$, i.e.,

$$\nu_{max} = 2NA / \lambda \qquad (15)$$

We may also note that at this limiting condition, only the +1 order is collected, i.e., not all of the diffracted light can be collected, so that the modulation is already low. The MTF curve can be described mathematically in terms of the above variables. As an example, the MTF $[(H(\nu)]$ of a lens employing incoherent illumination and a circular exit pupil is given by

$$H(\nu) = \frac{2}{\pi} [\cos^{-1}(\nu/\nu_{max}) - \nu/\nu_{max}\sqrt{1-(\nu/\nu_{max})^2}] \qquad (16)$$

where ν_{max} is $2NA/\lambda$ (from equation 15).

The MTF vs. spatial frequency curves (see Figure 18) provide a convenient description of the variation of image quality with spatial frequency since the higher the value of the MTF for a given frequency, the greater the contrast of the projected image. Figure 18 shows the modulation M of the image intensity as a function of spatial frequency ν for a sinusoidal amplitude transmission grating for different values of the coherency factor, σ. We may note from this figure that the MTF of a fully coherent optical source has a spatial frequency cut-off one half that of a perfectly incoherent source as required by Equations 13 and 15. It is further apparent from Figure 18 that partial coherence will be beneficial for recording media which require an MTF > 0.4 as is typical of most organic photoresists.

It is generally considered that the best illumination for optimum pattern reproduction in conventional photoresists is a partially coherent source with a value of $\sigma \approx 0.7$. Values of $\sigma < 0.7$ result in higher values of the modulation transfer function for a given frequency; however, increasing the coherence is achieved by reducing the effective source size resulting in longer exposure time, i.e., less energy per total area delivered to the wafer. The image integrity of an isolated edge is also degraded for values of $\sigma <$

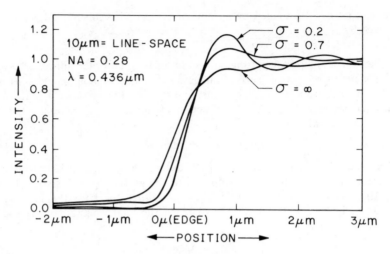

Figure 21. Image intensity profile at the edge of an opaque line in a grating of 10 μm lines and spaces.

0.7; this is caused by a phenomenon known as *ringing*, as illustrated for $\sigma = 0.2$ in Figure 21 (*21*). With coherency values $\sigma > 0.7$, it becomes increasingly difficult to maintain linewidth control since the intensity of the lower part of the intensity distribution curve increases with increasing σ. For values of $\sigma > 0.7$, there is also a decreased tolerance to focusing errors, i.e., a degradation of the depth of field of the optical system. From these considerations, the value of $\sigma = 0.7$ represents an acceptable tradeoff between these various factors.

A direct result of Equations 13 and 15 is that lenses with higher numerical apertures will result in higher resolution capabilities for an optical system. Figure 22 shows the modulation for a 1X lens and a 10X reduction lens, using radiation of 436 nm wavelength. It is clear from this figure that the 10X reduction lens provides a modulation transfer capability larger than that of the one-to-one projection lens. This has nothing to do with the reduction ratio per se but is a direct result of the higher numerical aperture of the 10X lens, although it is achieved only at the expense of image field (the image field diameter of a 10X reduction lens is ten times smaller than that for the 1X lens). Thus, higher resolution via higher *NA* is obtained at the expense of image field diameter and imposes an important constraint on the hardware design. Current state-of-the-art refractive lens design permits the placement of $\approx 10^8$ resolved elements (or spots) within an image field. With refractive lenses, this number depends little on the field size or resolution. Thus, for a given numerical aperture and wavelength, the limiting resolution will be obtained over a square field approximately given by 10^8

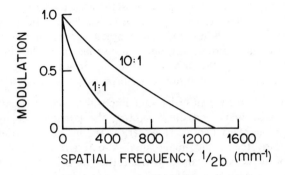

Figure 22. Modulation for a 1:1 lens (NA = 0.125) and a 10:1 reduction lens (NA = 0.33) using a wavelength of 436 nm. (Reproduced with Permission from Ref. 1)

divided by the spatial frequency squared. As an example, a system operating at a particular wavelength with a lens whose numerical aperture permits usable resolution of 1 - μm lines and spaces could exhibit such resolution only over a square field of approximately 2 cm. on a side. Higher resolution can be obtained only at the expense of field size in a manner that usually results in fewer total resolution elements within the imaged field. Such a reduction in field size requires a step-and-repeat approach to pattern an entire wafer.

2.1.e Wavelength Considerations.   It will be recalled from Equations 1 and 3 in Sections 2.1.a and 2.1.c respectively, that the minimum resolvable feature size is directly proportional to the wavelength of the exposing radiation. The resolution of contact and proximity printers shows a *one-half* order dependence on wavelength (Equation 1), whereas the wavelength dependence is *first* order for refractive or reflective projection systems (Equation 3). Hence it should be possible to improve the resolution of both techniques by decreasing the wavelength. Refractive lens systems typically use monochromatic light with wavelengths around 400 nm. Unfortunately, difficulties in the fabrication of refractive lenses for short wavelength applications have up until now, precluded these systems from consideration. The same is not true of *reflecting* optical systems or shadow printing techniques where resolution can in principle be increased quite simply by reducing the wavelength, provided that the aberrations are suitably low at the reduced wavelength. For example, Bruning (30) has demonstrated a resolution improvement using 310 - nm radiation in conjunction with a modified optical design for a one-to-one Perkin-Elmer system.

Conventional, positive photoresists can be used for wavelengths down to $\approx$ 300 nm. However, most of these resists have optical densities > 1 at wavelengths below 300 nm and as a consequence, are not useful at such low wavelengths since most of the light is absorbed near the surface with very little light reaching the resist/substrate interface. If we are to operate below 300 nm, new resist systems and processes must be designed.

2.1.f Depth of Focus. In order that line-width control be maintained, it is necessary, to a first approximation, that the latent image remain in focus throughout the resist depth. However, as the image plane departs from the plane of best focus, the image quality deteriorates. There is a certain amount of defocus tolerance wherein the image will still remain within specifications. Although influenced to some extent by developing conditions, this defocus tolerance, called the depth of focus, δ, may be derived from geometrical considerations bearing in mind the Rayleigh criterion for focus. According to Rayleigh, an optical path difference of $\pm$ $\lambda/4$ in the image plane does not seriously affect image quality in which case δ is given by $\lambda/2(NA)^2$. We can immediately see that the depth of focus is proportional to the inverse of the square of NA. Hence, although resolution may be improved by increasing NA, it is only at the expense of depth of focus. Thus a compromise must generally be reached.

2.1.g Standing Wave Effects. The projection of a perfect image onto a resist surface will not necessarily result in the replication of that image into a thick organic resist. When polymeric photoresist films are exposed with monochromatic radiation, standing waves are formed in the resist as a result of coherent interference from reflecting substrates creating a periodic intensity distribution in the direction perpendicular to the plane of the resist.

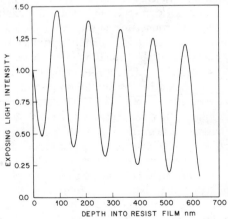

Figure 23. Plot of exposing light intensity as a function of depth in a photoresist on a reflecting substrate. (Reproduced with permission from Ref. 31)

The intensity distribution has a period related to the exposing wavelength. Dill et al. *(31,32)* have comprehensively treated this phenomenon both theoretically and experimentally. Their results are depicted in Figure 23 which illustrates the image intensity distribution as a function of depth in the resist. The light intensity decreases with increasing depth in the resist because of optical absorption. It should be noted that this intensity pattern changes during exposure because of "bleaching" of the light-absorbing molecules in the resist. Although the magnitude of the interference maxima and minima changes, their positions do not. Exposure under this standing wave condition results in variations in light intensity perpendicular to the resist film by as much as a factor of 3 from one position to another. In the case of a positive photoresist, this periodicity results in a variation in the rate of development along the edges of a feature leading to contours in the developed image as seen in Figure 24. The experimentally observed fringes correspond closely with those calculated by Dill, Neureuther and others as shown in Figure 25. It is obvious that this fringe structure deteriorates resolution and represents a serious limitation for small features. This effect is further complicated since its severity is related to the reflectivity of the substrate (which can vary), resist thickness and optical density of the resist.

An additional complication associated with the standing wave effect occurs for the exposure of resist over topography. When a resist is spin-coated onto a substrate containing steps, the resist thickness varies from one area to another on the wafer. Since the standing wave effect is a strong function of resist thickness, exposure variations resulting from variation in resist thickness in the vicinity of the step result in changes in linewidth.

Several approaches can be taken to minimize the effect of standing waves. Since the phenomenon results from coherent interference of mono-chromatic photons it can be reduced by employing broadband illumination.

Figure 24. Electron micrographs of 1.2 μm lines in a positive photoresist exposed on a reflective substrate and developed in standard photoresist developer. (Reproduced with permission from Ref. 32)

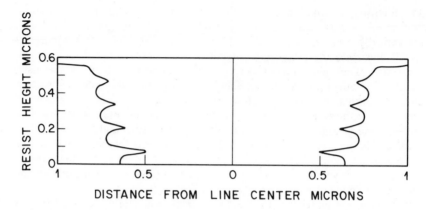

Figure 25. Calculated edge profiles for a nominal 1.0 μm line in positive photoresist. (Reproduced with permission from Ref. 1)

However, this approach introduces additional complications in optical systems, requiring complex lenses to correct for chromatic aberrations. Alternatively, the effect can be minimized by optimizing the resist thickness and substrate reflectivity (3) or by using post-exposure baking schedules designed to smooth the edge by flowing the polymeric material (33). The use of concentrated developers or overdeveloping also reduces the standing wave effect; however, both baking and overdeveloping merely improve the edge of the feature and do not eliminate the sensitivity of line size over topography.

Another approach to reducing this problem involves depositing a thick polymeric layer on the wafer which planarizes the surface. This planarizing polymeric film is then coated with a photoactive polymer which is patterned. This structure permits the image to be recorded in a thin resist on a planar reflectionless substrate. The image is then reactively ion etched into the underlying thick film resulting in minimum linewidth variation over the topography. This subject will be dealt with extensively in Chapter 6.

After consideration of all factors which limit resolution such as exposure hardware, resist systems, registration, alignment, and linewidth control, there is a general consensus that the useful resolution limit of photolithography may lie somewhere between 0.4 and 0.8 μm and depends on such factors as the implementation of short wavelength UV and the ability to accurately place images from a projection tool onto a silicon wafer.

2.2 Electron Beam Lithography

It has been recognized for many years that a finely focused beam of electrons can be deflected accurately and precisely over a surface. If this surface is coated with a radiation-sensitive polymeric material, the electron beam can be used to write patterns of very high resolution (*34-38*). Electron beams can be focused to a few nanometers in diameter and rapidly deflected either electromagnetically or electrostatically. Structures have been made with dimensions of less than 100 nm using this technique. Electrons, like photons, possess particle and wave properties; however, their wavelength is on the order of a few tenths of an angstrom, and therefore their resolution is not limited by diffraction considerations. Consequently the minimum linewidth that can be produced with electron beam exposure is much less than that achieved with photolithography. Resolution is limited, however, by forward scattering of the electrons in the resist layer and backscattering from the underlying substrate as we shall now examine.

2.2.a Electron Scattering. When an electron beam enters a polymer film (or indeed any solid material), it loses energy via elastic and inelastic collisions known collectively as electron scattering. Elastic collisions result only in a change of direction of the electrons while inelastic collisions result in energy loss. These scattering processes lead to a broadening of the beam, i.e., the electrons spread out as they penetrate the solid producing a transverse or lateral electron flux $J(r,z)$ normal to the incident beam direction (see Figure 26), and cause exposure of the resist at points remote from the point of initial electron incidence, which in turn results in developed resist images wider than expected. There are two types of scattering that are important, viz., forward and backscattering. Since most of the electrons are scattered in the forward direction through small angles ($< 90°$) relative to their original direction, this effect merely broadens the incident beam. Some electrons experience large angle scattering (approaching 180°) which causes these electrons to return to the surface. This situation is depicted in Figure 27 which shows the Monte Carlo simulated trajectories for 100 electrons projected on to the x-z plane for a 10- and 20 - kV point source. The figure qualitatively shows the degree of forward and backscattering. The forward scattered electrons are difficult to identify because of their high density and small lateral spread. On the other hand, the backscattered electrons are clearly evident being spread out over distances on the order of 1 μm for 10- kV electrons while at 20 kV the distance is 3-4 μm.

The magnitude of the electron flux at any point $J(r,z)$ depends upon such parameters as the atomic number and density of both the resist and substrate as well as the velocity (accelerating voltage) of the electron. The resultant undesired exposure of resist in areas not directly addressed by the electron beam is called the *proximity effect* and imposes certain restrictions

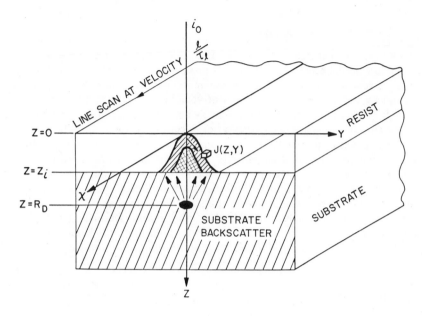

Figure 26. Three-dimensional electron scattering model for a resist on a thick substrate with a scanning electron beam of zero diameter.

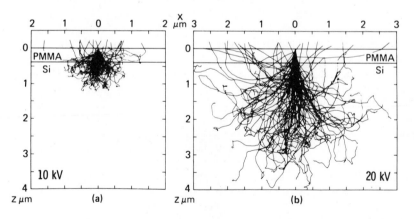

Figure 27. Monte Carlo simulated trajectories of 100 electrons in PMMA resist on silicon (Reproduced with permission from Ref. 42)

on the size and shape of relief structures which can be achieved in the resist, particularly for complex patterns with high packing densities and dimensions < 1 μm. Although it is possible to fabricate structures with dimensions approaching tens of nanometers, a thorough understanding of the factors that affect the shape and size of the developed relief structure after exposure is important to the practical application of electron beam lithography. These factors have been the subject of extensive theoretical and experimental research. In particular, several models have been developed that allow the prediction of exposed resist profiles and that in turn allow pattern shapes and/or exposure conditions to be tailored in such a way as to make the exposed patterns conform as closely as possible to the desired pattern. Such models are also very useful in understanding the effects of beam energy, incident dose, substrate material and pattern geometry (proximity effects).

The fundamental parameters necessary to determine profile shape are the absorbed energy density and its dependence on spatial position within the resist film. The situation is depicted in Figure 28 where the absorbed energy density $E(r,z)$ within a given volume element at point r,z resulting

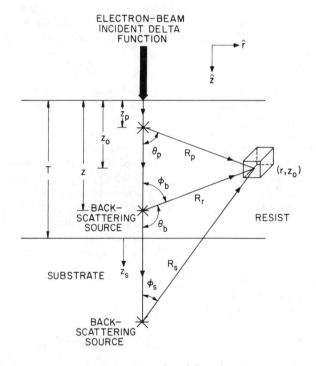

Figure 28. Geometry for computing the delta function response of the resist. (Reproduced with permission from Ref. 43)

from an incident delta function (beam of zero diameter) is seen to contain contributions from scattering within the resist (forward and backscattering) and from within the substrate (backscattering). For the simplest case of a stationary electron beam with current i turned on for a time interval τ we may write (39)

$$(i\ \tau/e)\ I(r,z) = E(r,z)eV/cm^3 \qquad (17)$$

where $E(r,z)$ is the energy dissipated per init volume, (eV/cm^3), $i\ \tau/e$ the incident number of electrons, and $I(r,z)$ the energy dissipated per unit volume per electron $(ev/cm^3/electron)$. Thus the absorbed energy density is the product of the electron flux passing through an elemental volume surrounding the observation point and the energy loss in that elemental volume for each electron. $I(r,z)$ is of fundamental significance since it allows one to calculate the energy dissipated per unit volume at any point in the polymer film for any given exposure pattern. The calculation of both $I(r,z)$ and the electron flux is the primary objective of various exposure models which are designed to examine the effect of exposure variables on profile shape.

The prediction of a given contour shape, i.e., resist profile, requires both exposure *and* development models. We will first examine the various exposure models which have been developed and then combine these with development studies in order to predict resist profiles and compare them with experiment.

2.2.a.1 Monte Carlo Methods. The Monte Carlo method attempts to simulate via a digital computer the trajectories of the incident electrons within the resist/substrate system. This involves following the electron through a succession of distinct scattering events during which it undergoes angular deflection and energy loss. Most approaches employ a single scattering model in which the direction of scattering is chosen by a random number that is weighted by the screened Rutherford expression for the scattering cross-section. In this process, the angular deflections suffered by the incident electrons are assumed to occur via elastic scattering with the target nuclei with the angle of scattering being calculated for each event in accordance with the differential scattering cross-section for a nucleus in a screened atomic field. The differential scattering cross-section is calculated using the Born approximation to the Schröedinger equation (40,41). The angular distribution of scattered electrons is dependent on the assumed potential $V(r)$. Most calculations employ the Thomas-Fermi Potential which assumes that an incoming electron sees the atomic charge of the nucleus screened by the electron cloud of the atom. The potential has the form

$$V(r) = -\ Ze^2/r\ \exp[-0.745\ r\ Z^{1/3}/a_o]$$

where a_o is the Bohr radius equal to 0.53×10^{-8} cm. Using this atomic potential, the differential scattering cross-section per unit solid angle is given by

$$\frac{d\sigma}{d\Omega} = \frac{Z_i(Z_i + 1)e^4}{4m^2v^4[\sin^2(\theta/2) + \alpha_i^2]^2} \tag{18}$$

where m is the mass of the electron, v the velocity of the electron, and α_i the atomic screening parameter given by

$$\alpha_i = 2.33\, Z_i^{1/3}\, E^{-\frac{1}{2}} \tag{19}$$

Between elastic scattering events, the electrons are assumed to travel in straight lines (of length equal to the mean free path) undergoing energy loss. This energy loss is usually modeled via the continuous slowing down approximation (CSDA) in which the electron loses energy continuously along each free flight segment according to the Bethe energy loss formula

$$\frac{dE}{dx} = -\frac{2\pi e^4 n_e}{E} \ln\left[\frac{aE}{I}\right] \tag{20}$$

where n_e is the density of atomic electrons, I the mean excitation energy, E the energy of the incident electron, and a a constant equal to 1.166. Other models have modified the CSDA approach to take account of the statistical nature of inelastic collisions. Given a finite step length, an incident electron can in principle suffer an amount of energy loss ranging from zero up to its initial energy at the beginning of the step. The Bethe stopping power formula pertains only to the mean or average rate of energy loss with no account being taken of the statistical fluctuations in the energy loss process. In practice, these refined models produce results which are not significantly different from the CSDA approach.

The sequence of events for one electron is shown in Figure 29 for an electron with energy E_o impinged at the origin (position 0) and at 90° to the surface of a semi-infinite target (*42*). The first scattering event is assumed to occur at the origin. The scattering angles θ_o (conical angle of scattering in the Rutherford expression) and ϕ (azimuthal angle of scattering) and the step length Λ_o are calculated by the Monte Carlo techniques; then the spatial position 1 of the next scattering event is determined with respect to 0. Within the step length, the electron is assumed to have constant energy E_o. The electron energy at point 1 is then calculated by decrementing the energy with respect to its value at point 0 via the Bethe expression for energy loss per unit distance. At point 1, the sequence is repeated using E_1 to calculate θ_1, ϕ_1, and Λ_1. The sequence is repeated continuously until the energy has degraded to some arbitrary value close to the mean ionization energy. By carrying out many such simulations, trajectory patterns as those shown in Figure 27 can be generated and an absorbed energy density matrix can be calculated from which I(r,z) at any point can be determined.

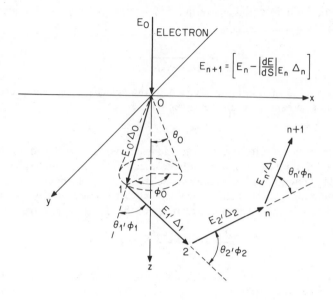

Figure 29. Geometry for Monte Carlo simulation in thick targets.

2.2.a.2 Analytical Models. In the Monte Carlo technique, it is necessary to calculate the trajectories of several thousand electrons in order to obtain statistically meaningful results. This implies a great deal of computer time and expenditure. Thus for practical as well as heuristic reasons, analytical models to predict energy dissipation are useful. In these models, the energy dissipated by an electron beam is considered to be composed of three contributions: (i) small angle forward scattering of the incident beam in the polymer, (ii) large angle scattering from the substrate, and (iii) large angle backscattering of the incident beam within the polymer.

   Two analytical models have been used to determine the forward spreading of the incident electrons. Greeneich and Van Duzer (*43*) based their model on Lenz's plural scattering theory in which the angular distribution of scattered electrons is obtained by integrating the Boltzmann equation over all space. A more simple approach makes use of the simplified Boltzmann Transport Equation in the small angle multiple approximation originally derived by Fermi (*40*). This multiple scattering model assumes that the spatial distribution of the electron flux is due to many small angle collisions and neglects the change in electron energy in evaluating the spatial distribution. The treatment is thus pseudo-monoenergetic and is restricted to thin resist films where the average energy of electrons crossing the interface is equal to the incident energy. This model evaluates directly the spatial distribution of the incident electrons within the resist in terms of a probability function that expresses the probability that an electron will be scattered to a radial distance r after penetrating a vertical distance z. The probability expression has the form

$$H(r,z) = \frac{3\lambda}{4\pi z^3} \exp(-3\lambda r^2/4z^3) \tag{21}$$

where r is the radial distance from the axis of the incoming electron, z the vertical distance along the axis of electron penetration and λ the transport mean-free path which in the nonrelativistic approximation is given by (40)

$$\lambda(\text{Å}) = \frac{5.12 \times 10^{-3} E^2 A}{\rho z^2 \ln(0.725 E^{1/2} Z^{-1/3})} \tag{22}$$

where A is the mass number, E is the electron energy, ρ is the density and Z is the atomic number. Equation 21 shows the spatial distribution of electrons is Gaussian at any depth for a delta input function. When the impinging electrons have a Gaussian distribution, i.e. ,

$$J(r) = i_o/\pi \bar{r}^2 \exp-(r^2/\bar{r}^2) \tag{23}$$

where i_o is the beam current and $\bar{r}$ the Gaussian radius (root-mean-square radius), then the resulting density distribution inside the solid will be given by the convolution of the integral of the input function $J(r)$ and the response function $H(r,z)$ as

$$J(r,z) = \int_0^{2\pi} \int_0^\pi H(r,z) J(r) r \, dr \, d\theta \tag{24}$$

This integration cannot be carried out rigorously in closed analytical form and thus requires numerical integration. However, a good approximation is obtained using the approximation that in a Gaussian distribution the value of $\bar{r}$ is the root-mean-square uncertainty in the radial position of an electron. If scattering now occurs, an additional uncertainty is introduced [in this case $(4z^3/3\lambda)^{1/2}$], and if the two uncertainties are random or independent, they can be combined in quadrature to yield the minimum dispersion of the Gaussian (44). This dispersion $\bar{s}$ is the minimum uncertainty in electron radial position and is simply

$$\bar{s} = (\bar{r}^2 + 4z^3/3\lambda)^{1/2} \tag{25}$$

for a spreading Gaussian profile. The Gaussian profile of the beam is therefore maintained as it penetrates the resist and is given by

$$J(r) = \frac{i_o}{\pi \bar{s}^2} \exp{-(r^2/\bar{s}^2)} \tag{26}$$

The energy loss per cm^3 per electron caused by the forward scattered electrons can then be calculated as

$$I(r,z)\big|_{FR} = \frac{dE}{dx} \times \frac{J(r)}{i_o} \tag{27}$$

When the incident electron passes close enough to an atomic nucleus in the resist or substrate, it can be scattered through angles greater than 90° relative to the incident direction. The calculation of $I(r,z)\big|_{BS}$ and $I(r,z)\big|_{BR}$ resulting from such large angle backscattering from the substrate and resist follows the approach of Everhart (41,45) which assumes that electrons undergo only one large-angle Rutherford scattering event coupled with the Thomas-Widdington law to define an electron range in the substrate. The reflection coefficient is calculated and a parameter in the model adjusted to obtain a good fit to experimental reflection coefficient data for atomic numbers $Z < 50$. The absorbed energy density per electron is then obtained as the sum of the individual contributions, i.e.,

$$I(r,z) = I(r,z)\big|_{FR} + I(r,z)\big|_{BS} + I(r,z)\big|_{BR} \tag{28}$$

and the total absorbed energy density is obtained as the product of $I(r,z)$ and the total electron flux.

The calculation of backscatter coefficients via the approach outlined above is mathematically complex. Heidenreich (44) developed a simple empirical backscatter model which is applicable to resist exposure being based on the direct observation of chemical changes produced by backscattered electrons at different accelerating voltages on several substrates. The model is independent of scattering trajectory and energy dissipation calculations and is essentially a radial exponential decay of backscatter current density out to the backscatter radius determined by electron range.

An alternative empirical approach for estimating effects of backscatter involves exposure of effectively free standing thin resists. The model (46) is most conveniently discussed with reference to the well-known contrast curve of a negative resist shown in Figure 30 for COP exposed to 20-kV electrons on three different substrates as explained in the figure and its accompanying caption. The curve corresponding to a layer of resist on 100-nm Si_3N_4 represents a good approximation to the ideal case of a free-standing resist film because the backscatter coefficient from such a substrate is very small being on the order of 0.01-0.02.

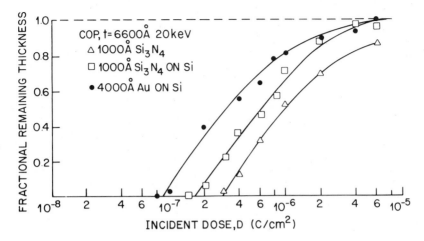

Figure 30. Exposure curves for COP negative electron resist on three substrate materials.

For uniform large area exposure such as one encounters in the determination of the contrast curve, the incident electron energy dissipation density depends only on the depth z into the resist. Explicitly, it can be written in the form (*47,48*)

$$E(z) = \frac{D}{e} \left[\frac{dV}{dz} \right]_z \qquad (29)$$

where D is the incident areal input dose (C/cm^2), e is the electronic charge and $\frac{dV}{dz}$ is the energy dissipation per unit distance in the z direction. The specific dissipation can be written as

$$\frac{dV}{dz} = \frac{V_a}{R_G} \Lambda(f) \ eV/cm \qquad (30)$$

for beam voltage V_a where R_G is the Grün range given by

$$R_G = (0.046/\rho) \ V_a^{1.75} \qquad (31)$$

and $\Lambda(f)$ the depth-dose function expressed in terms of normalized penetration ($f = z/R_G$). The density ρ is approximately unity for polymeric resists. The depth-dose function is an empirical quantity based on measurement of ionization as a function of beam penetration or a measurement of penetration of incident electrons as a function of accelerating potential. It is expressed as a polynomial of the form (49).

$$\Lambda(f) = 0.74 + 4.7f + 8.9f^2 + 3.5f^3 \tag{32}$$

Figure 31 shows a plot of $\Lambda(f)$ vs. f where we see that the maximum energy dissipation rate occurs at some finite distance from the interface and is dependent on the accelerating voltage.

We may consider $\dfrac{dV}{dz}$ to be the sum of two terms, viz.,

$$\frac{dV}{dz} = \left|\frac{dV}{dz}\right|_{FS} + \left|\frac{dV}{dz}\right|_{BS} \tag{33}$$

where $\left|\dfrac{dV}{dz}\right|_{FS}$ is the energy dissipation caused by forward scattering and $\left|\dfrac{dV}{dz}\right|_{BS}$ is the energy dissipation caused by backscattering. Equation 33

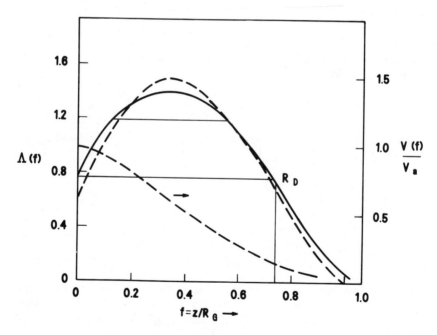

Figure 31 Depth-dose function for electrons in a target of low atomic number. The normalized penetration $f = z/R_G$.

can be rewritten as

$$\frac{dV}{dz} = \left|\frac{dV}{dz}\right|_{FS} \left\{1 + \left|\frac{dV}{dz}\right|_{BS} \Big/ \left|\frac{dV}{dz}\right|_{FS}\right\} \tag{34}$$

$$= \left|\frac{dV}{dz}\right|_{FS} \{1 + \eta\} \tag{35}$$

where η is defined as the backscatter coefficient. For the case of the thin Si_3N_4 substrate, $\eta \approx 0$ so that from Equation 29 we have

$$E_g^i = \frac{(D_g^i)_o}{e} \left(\frac{dV}{dz}\right)_{FS}^i \tag{36}$$

where D_g^i is the incident dose necessary to form gel (referred to as the interface gel dose) and E_g^i is the energy absorbed at the interface. For any other substrate

$$E_g^i = \frac{D_g^i}{e} \left\{ \left(\frac{dV}{dz}\right)_{FS}^i + \left(\frac{dV}{dz}\right)_{BS}^i \right\} \tag{37}$$

Thus from Equations 36 and 37 we obtain

$$(D_g^i)_o / D_g^i = 1 + \eta^i \tag{38}$$

where η^i is the effective backscatter coefficient evaluated at the interface.

The depth-dose model allows the gel energy E_g^i of a resist to be calculated from an observed interface gel dose D_g^i under conditions of zero backscatter and predicts a dependence of D_g^i on accelerating voltage going as $V_a^{-0.75}$ which agrees reasonably well with experiments. Calculations of backscatter coefficients are in good agreement with Monte Carlo calculations (46). Further, backscatter contributions to the electron flux at any point calculated from the multiple scattering model may be simply accommodated by modifying the forward scattered beam density as $J_o(1 + \eta)$.

Comparison of results obtained from analytic models with Monte Carlo calculation are in good qualitative agreement and several generalizations may be made. The energy dissipation can be visualized as arising from two sources: the forward scattered incident electrons and the backscattered electrons from the substrate. The energy dissipated by the forward scattered electrons is localized in the vicinity of the beam axis. As the electrons travel through the resist, they scatter laterally, and the region of exposure by the forward scattered electron is increased at the resist-substrate interface compared with near the top of the film. The region over which the backscattered electrons expose the resist is defined by the electron range in the resist and the substrate. While the maximum energy dissipated per unit volume by the backscattered electrons is much less than that by the forward scattered electrons, the volume integrated contributions are comparable.

Exposure of the resist by the forward and backscattered electrons depends upon the beam energy, film thickness and substrate atomic number. As the beam energy increases, the energy loss per unit path length and scattering cross-sections decrease. Thus, the lateral transport of the forward scattered electrons and the energy dissipated per electron decrease while the lateral extent of the backscattered electrons increases due to the increased electron range. As the resist film thickness increases, the cumulative effect of the small angle collisions by the forward scattered electrons increases. Thus the area exposed by the scattered electrons at the resist-substrate interface is larger in thick films than in thin films. Proper exposure requires that the electron range in the polymer film be greater than the film thickness in order to ensure exposure of the resist at the interface. As the substrate atomic number increases, the electron reflection coefficient increases which in turn increases the backscattered contribution. In addition since the electron range in the substrate also decreases with increasing atomic number, the exposure by the backscatter electrons is concentrated near the beam axis.

2.2.a.3 Development Modeling. In order to predict resist profiles, development models are required coupled with the exposure models discussed in the previous sections. The exposure of positive polymeric resists which operate by main chain scission produces molecules having molecular weight M_f less than the original molecular weight M_n. The extent to which the molecular weight is reduced can be expressed in terms of the absorbed energy density $E (eV/cm^3)$ by (50,51)

$$M_f = \frac{M_n}{1 + \dfrac{G(S)\, EM_n}{100\, \rho\, A_o}} \qquad (39)$$

where ρ is the resist density, A_o is Avogadro's Number and $G(S)$ is the G value for scission (number of scission's per 100 eV of absorbed energy). Thus developed resist contours may be considered as representing contours of constant absorbed energy density. For example, if it is known that a given developer will dissolve polymer molecules with M_f less than some critical value, then the developed profile must represent the contour of absorbed energy corresponding to E_{crit}. This is the so-called *threshold energy density* model (43) and is applicable at high exposure densities to high contrast resists. Under such circumstances the developed resist profile will be time independent. Figure 32 shows a comparison of developed profiles for line doses of 0.3×10^{-8} and 1.0×10^{-8} C/cm for PMMA on an aluminum substrate at different accelerating voltages. Comparison of the theoretical contours calculated from the model of Greeneich and Van Duzer with the empirical contours of Wolf et al. (52) shows reasonable agreement except at

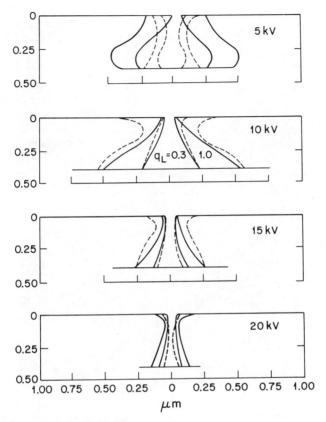

Figure 32. Comparison of developed profiles for charges per unit length of 0.3 × 10⁻⁸ C/cm and 1.0 × 10⁻⁸ C/cm for PMMA on an aluminum substrate. The solid lines are from theory; the broken lines are the experimental contours of Wolf et. al. (52)

5 kV. Greeneich and Van Duzer attribute this to a lack of validity of the plural scattering theory for primary electrons at 5 kV. It is apparent from this figure that the developed contour depends on the beam energy raising the possibility that the contour can be tailored to a specific shape by proper choice of exposure parameters. This is demonstrated in Figure 33 for PMMA exposed at varying doses (53).

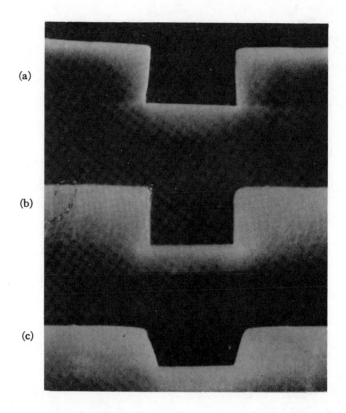

(a)

(b)

(c)

Figure 33. Actual PMMA resist profiles at an incident charge density of (a) 10^{-4} C/cm² (b) 8×10^{-5} C/cm², and (c) 5×10^{-5} C/cm² (Reproduced with permission from Ref. 53)

As pointed out by Hatzakis et al. (54) and by Greeneich (51), the threshold energy density model is not adequate for time-dependent developers with low contrast resists. Hatzakis et al. obtained contour lines for equal absorbed energy in the resist and expressed it as an energy matrix $E(r,z)$ which was determined from the product of the electron intensity distribution and the energy dissipation function. Knowing the change in molecular weight for a given exposure level and the dependence of solubility rate on fragmented molecular weight, they were able to transform the two-dimensional energy density matrix into a solubility rate matrix. Greeneich used a similar model based on a more rigorous analysis of absorbed energy density. He used an empirically determined formula for solubility rate in the form

$$R = R_0 + \frac{\beta}{M_f^{\alpha}} \qquad (40)$$

where R_0, β, and α are constants for a given developer. A typical time-developed profile in PMMA observed for a low incident dose is shown in Figure 34.

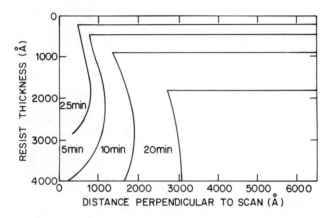

Figure 34. Time evolution of developed contours in PMMA developed in MIBK. (Reproduced with permission from Ref. 51)

The bulk of profile modeling work has been carried out using PMMA as the resist primarily because the development process for this material can be modeled as a surface-limited etching reaction controlled by the local energy dissipation density. On the other hand, the significant amount of swelling that accompanies the removal of uncrosslinked polymer during development of a negative resist is much more difficult to treat. Heidenreich et al. (*48*) have proposed a phenomenological model to simulate negative resist profiles which, while not taking swelling into account, does produce results that are in qualitative agreement with experiment. Starting with the simplified solution to the multiple scattering model for predicting electron density distributions as given in Equation 21 and incorporating backscatter as a modification of the forward current, Heidenreich obtained the following expression for the dose $D(y,z)$ at distance y from the line center and at penetration z for a beam scanning along the x direction at line speed $\dot{x}$:

$$D(y,z) = \frac{(1 + \eta) D_x^o}{\bar{s}(z) \sqrt{\pi}} \exp-(y^2/s(z)^2) \tag{41}$$

Here, η is the backscatter coefficient for the substrate and D_x^o the line input dose which is equal to $i_o/\dot{x}$. The Gaussian beam radius in the resist at penetration z is

$$\bar{s}(z) = (\bar{r}^2 + \frac{4z^3}{3R_B})^{1/2} \tag{42}$$

where R_B is the Bethe range (cf. Equation 25 only here the diffusion length λ is taken at the Bethe range R_B).

The development action in negative resists consists of leaching soluble polymer from the irradiated region. The remaining crosslinked gel then collapses to produce the final line. Provided the line maintains adhesion at the substrate, the line width will not be changed on shrinking. The line height can depend upon the two dimensional shrinkage mechanism, but comparison with experiment indicates the height agrees with the center line dose ($y=0$) for lines wider than about 0.4 μm. The areal input dose at the line center is given by

$$D(0) = \frac{(1 + \eta) D_x^o}{\bar{r}\sqrt{\pi}} \; C/cm^2 \tag{43}$$

By convolving the spreading Gaussian beam treatment with the experimentally determined relationship between resist thickness and input dose, Heidenreich was able to calculate line heights and profile widths at the substrate in good agreement with experiment.

Chung used a similar approach to model negative resist profiles. For a resist on very thin low atomic number substrates such as Si_3N_4, $|\frac{dE}{dz}|_{FS}$ is essentially constant in z so that the measured contrast curve can be converted into a response curve connecting the fractional thickness remaining p with the energy dissipated in the film rather than incident charge density as is usually done. This gives us the basic response function of the particular resist as

$$p = F(E) \tag{44}$$

Chung extended this treatment to spatially varying energy dissipation densities as occurs in a line exposure. Assuming the x-axis is parallel to the direction of scan, the energy dissipation density will be a function of the coordinates (y,z) as discussed previously. Chung determined values of $E(y,z)$ directly from Monte Carlo calculations by convolving the δ-line response with the appropriate Gaussian beam. At any given value of y, the resist thickness z_o is divided into a number of segments of height Δz_i as depicted in Figure 35. For sufficiently small values of Δz_i, the energy dissipation $E(y, z_i)$ can be regarded as constant throughout. Thus each infinitessimal segment Δz_i will give rise to an infinitessimal remaining thickness $\Delta z_i'$ determined by $E(y, z_i)$ from the modified contrast curve. The total thickness remaining after development at y is the sum

$$p(y) = \frac{z_y}{z_o} = \underset{\Delta z_i \to 0}{\underset{N \to \infty}{Limit}} \frac{\sum_{i=1}^{N} \Delta z_i'}{z_o} \tag{45}$$

$$= \frac{1}{z_o} \int_0^{z_o} F(E(y, z_i)) dz \tag{46}$$

where $F(E(y, z_i))$ is the fractional thickness remaining (obtained from the modified contrast curve) for a given value of absorbed energy density. While this simple phenomenological approach completely ignores swelling effects and is essentially one dimensional in nature, the predictions derived from the model are in qualitative agreement with experiment.

By and large the theoretical framework for evaluating profile shapes is well established and has led to a better understanding of the choice of operating parameters. It has been particularly helpful in generating proximity effect algorithms in order to specify exposure requirements for complex geometries (see ref. 39 for a review of proximity correction procedures).

2.2.b Electron Lithography Systems and Limitations. Electron beam methods have been pursued since the late fifties for their high resolution capability. Following the successful development of scanning electron microscopes, it was soon recognized that modification of these systems would offer the capability of directly generating high resolution patterns with speed, and precision. The first *experimental* electron beam writing systems were designed to take advantage of the high resolution capabilities offered by electron beam writing (55,56). The first *commercial* hardware, however, was built to take advantage of attributes other than resolution, such as the direct pattern generation from computer software (mask fabrication) and economic custom interconnection of integrated logic circuits. Electron beam lithography has since become the dominant technology for making master masks for photolithography where important advantages in the area of image placement accuracy and line width control are realized (57).

Electron beam systems can be conveniently considered in two broad categories: those using scanned, focused electron beams which expose the wafer in serial fashion, and those projecting an entire pattern simultaneously, onto a wafer. Electron beam projection systems have been investigated extensively since they offer the potential of higher exposure rates as a

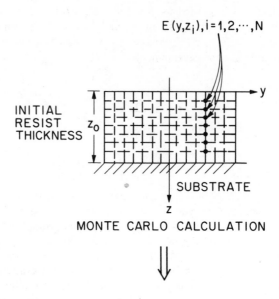

$E(y,z_i), i = 1, 2, \cdots, N$

INITIAL RESIST THICKNESS z_0

SUBSTRATE

MONTE CARLO CALCULATION

$$P(y) = \frac{z_y}{z_0} = \frac{\int_0^{z_0} F[\mathcal{E}(y,z)]\, dz}{z_0}$$

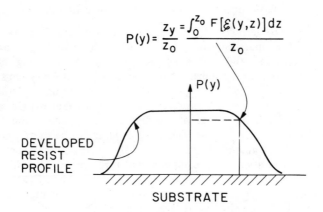

$P(y)$

DEVELOPED RESIST PROFILE

SUBSTRATE

SIMULATED RESIST PROFILE

Figure 35. Pictorial representation of simulation procedure.

result of projecting entire wafer images rather than serially exposing only small features or a portion of a feature at a time. Figure 36 illustrates a "family tree" of electron beam systems which includes both projection and scanning systems. Specific systems will be discussed in Sections 2.2.b.1 through 2.2.b.4.

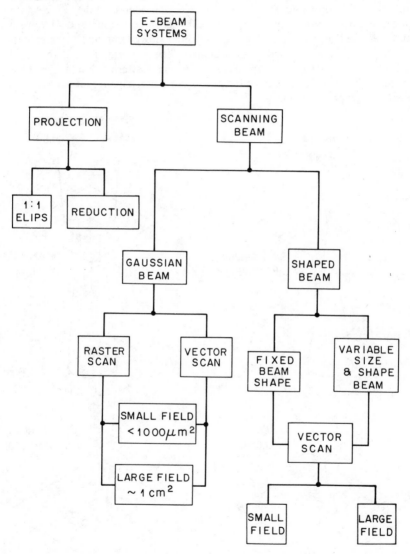

Figure 36. "Family tree" of electron beam systems.

Scanning beam systems can be subdivided into two categories: Gaussian, round beam systems and shaped beam systems shown in Figure 37. All scanning beam systems have certain hardware similarities that may be considered independently of the writing strategy used. Figure 38 shows a simplified schematic diagram of an electron beam exposure system which can be divided into four subsystems: 1) electron source (gun), 2) electron optical column (beam forming system), 3) mechanical stage and 4) control computer used to control the various machine subsystems and transfer pattern data to the beam deflection systems. The performance capability of each of these subsystems coupled with electron scattering limitations (discussed in Section 2.2.a) determine the overall performance of an electron beam lithographic system.

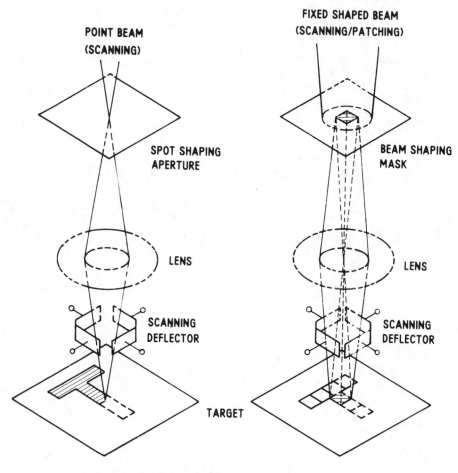

Figure 37. Schematic of round beam and shaped beam columns.

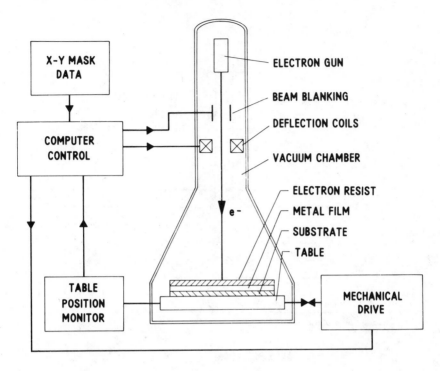

Figure 38. Schematic of an electron beam exposure system.

Figure 39 illustrates the exposure of a resist surface using a focused electron beam and is helpful in understanding the limitations on resist sensitivity imposed by the hardware. We consider in Figure 39 a single address point (area of the beam) with dimensions x and y and an area, a expressed in cm². An electron optical column can deliver within that area, some maximum current (i) which is limited primarily by the source brightness and column design. The computer that controls the electron beam limits the frequency at which the beam can be moved and modulated and results in a minimum "flash" time in seconds (t). From these considerations we can calculate the maximum dose (D_{max}) in Coulombs per sq cm (C cm⁻² that can be delivered by a particular electron beam exposure system by the following equation:

$$D_{max} = \frac{it}{a} \tag{47}$$

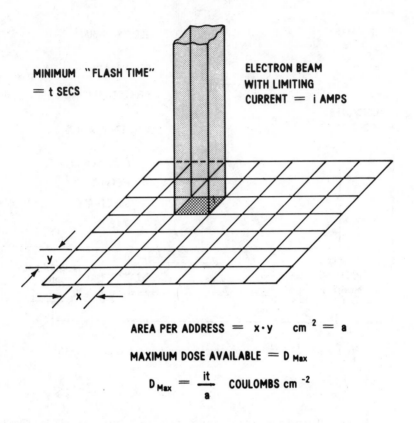

MINIMUM "FLASH TIME"
= t SECS

ELECTRON BEAM
WITH LIMITING
CURRENT = i AMPS

AREA PER ADDRESS = x·y   cm ^{2} = a

MAXIMUM DOSE AVAILABLE = D_{Max}

$$D_{Max} = \frac{it}{a} \quad COULOMBS \ cm^{-2}$$

Figure 39. Schematic of exposure of a unit address of resist surface.

This imposes a limitation on the sensitivity of the resist, i.e., it is necessary to design a resist that will react sufficiently to D_{max} to produce a lithographically useful, three-dimensional relief image. Now let us consider the four subsystems and their limitations.

Electron sources applicable to electron beam lithography are the same as those used in conventional electron microscopes. These sources can be divided into two groups-thermionic or field emission-depending on the way in which they emit electrons (58,59). Thermionic guns rely on the emission of electrons from a material that is heated above a critical temperature beyond which electrons are emitted from the surface. These sources are prepared from materials such as tungsten, thoriated tungsten, or lanthanum hexaboride, LaB_6. Tungsten emission sources have been used extensively in electron microscopes primarily because of advantages in current stability and tolerance to variations in vacuum conditions. In addition, they are easily fabricated and maintained. Tungsten sources provide a maximum brightness of $1-3 \times 10^5$ amps cm^{-2} steradian^{-1}.

Thoriated tungsten, whose surface has been carburized at high temperatures, have lower work functions than pure tungsten and emit equivalent electron beam currents at lower temperatures thereby extending the life and stability when compared to regular tungsten sources. These sources require a stable, high vacuum (10^{-7} to 10^{-8} torr) and are more difficult to fabricate.

Lanthanum hexaboride sources show brightness in excess of 10^6 amps cm^{-2} steradian^{-1} at 20 to 30 kV as a result of an even lower work function. They require a better vacuum than thoriated tungsten and are less stable. They are especially useful in illuminating shaped beam systems, since the emission is more uniform over a larger area.

Field emission sources use a high electric field surrounding a very sharp point of tungsten. The electric field extracts electrons at the tip of the source, forming a Gaussian spot of only a few tens of angstroms in diameter. These sources are considerably brighter (10^8 to 10^9 amps cm^{-2} steradian^{-1}) than any of the thermal emitters. These sources are more difficult to fabricate and must operate in vacuums on the order of 10^{-8} to 10^{-10} torr in order to achieve 1000-hour lifetimes. They are gaining acceptance as high brightness sources for scanning electron beam lithographic systems.

The electron optical column is the heart of any electron beam lithographic system and may be considered the active element in the system. An electron optical column consists of a source (described above), one or more lenses (used to focus and define the spot), a method of modulating the beam (beam blanker), beam limiting apertures, and a beam deflection unit used to position the beam precisely and accurately over the scan field. All of these elements can be arranged in a variety of ways depending on the purpose and design of the particular column and system (for a more detailed discussion, the reader is referred to references *60-62*). Suffice to say, the column is designed with a minimum of aberrations so that one can achieve a spot or an edge profile (for shaped beam systems) equivalent to the address structure of the particular electron beam exposure system. These aberrations can generally be reduced so as not to be the limiting factor for the desired resolution capability of a piece of hardware.

It is impossible to deflect an electron beam to cover an entire 3-to-4 inch diameter silicon wafer or mask substrate; hence mechanical stages are required to move the substrate through the deflection field of the electron beam column. Stages can be operated in a stepping mode (63) in which the stage is stopped, an area of the pattern written and then the stage moved to a new location where the adjacent pattern area is exposed. The correct stage location may be determined by the use of a laser interferometer or by locating registration marks on the substrate within each pattern area using the electron beam as a probe in a manner similar to that of a scanning electron microscope. Alternatively, stages can be operated in a continuous mode (64) where the pattern is written on the substrate while the stage is moving (see Figure 40). In order to position the pattern features accurately, one must know the stage location to a small fraction of a micron. Again, laser interferometers are used to monitor stage position continuously so that residual errors can be compensated by electronic deflection of the electron beam.

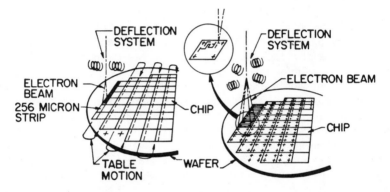

Figure 40. Operating modes for electron beam systems: left — raster scan coupled with continuous table motion; right — vector scan, step and repeat.

These complicated systems are controlled by means of a computer, which in real time monitors the various operating parameters of the column and stage position and which in addition, transfers primary pattern data directly to the electron deflection system. The rate at which the data can be transmitted to the electron optical column ultimately governs the modulation rate of the electron beam machine, i.e., flash time. Modulation rates in excess of 100 MHz have been achieved, and it is conceivable that much higher modulation rates will be attained in the future.

The preceding remarks on scanning electron beam systems are applicable to both round beam, raster scanning systems as well as round beam, shaped beam and variable shaped beam vector scanning systems. Raster and vector scanning represent the two basic scanning methods employed in electron lithographic systems (65,66) (see Figure 41)

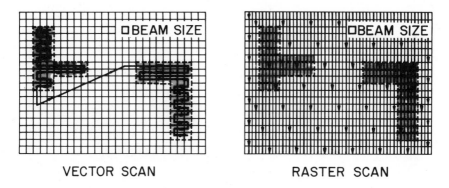

VECTOR SCAN   RASTER SCAN

Figure 41. Comparison of vector scan and raster scan writing schemes.

2.2.b.1 Vector Scan Systems.   With vector scanning, the electron beam is deflected sequentially to individual elements of the pattern rather than scanned serially over the entire wafer. Exposure time is saved when this approach is used for circuit levels whose total exposure area is less than 20% since no time is spent in areas not to be addressed by the beam. However, more complicated beam deflection systems must be designed in order to compensate for hysteresis effects and large angle deflections. Several successful machines have been built using this approach including the Philips beamwriter, the IBM VSl, the Texas Instruments EBSP and the Cambridge EBMF2. All of the above machines use a Gaussian round beam and step-and-repeat stage motion. The vector scanning methodology lends itself to the use of *shaped beams* and *variable* shaped beams instead of the more conventional round beams. In shaped beam systems an aperture of a given geometric configuration (usually rectangular or square) is imaged through electromagnetic lenses and projected onto a wafer surface (see Figure 37). This modification offers the advantage of simultaneously exposing several address units. A further modification of the shaped beam approach is that of imaging two apertures simultaneously resulting in the ability to vary the *size* and *shape* of the beam thus minimizing the number of flashes required to expose larger patterned areas (see Figure 42). It is also possible to project complex geometric characters with an electron beam system, and this approach is especially attractive for writing patterns of repeated constant geometric shapes (as in magnetic bubble devices or MOS memory structures). Figure 43 depicts the various pattern geometries and beam profiles for these approaches.

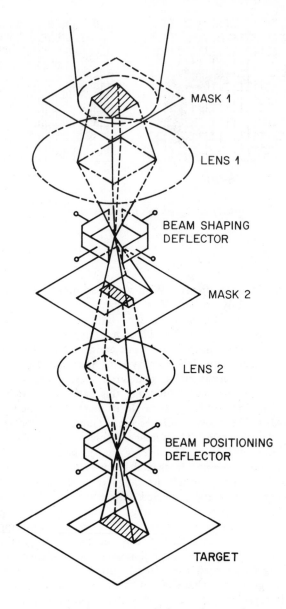

MASK 1

LENS 1

BEAM SHAPING
DEFLECTOR

MASK 2

LENS 2

BEAM POSITIONING
DEFLECTOR

TARGET

Figure 42. Diagram of a compound shape electron beam system

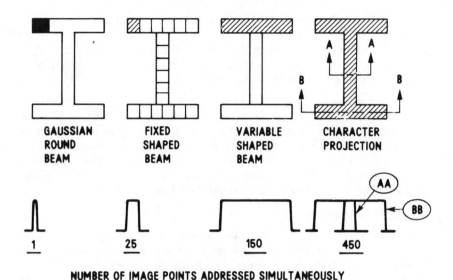

GAUSSIAN ROUND BEAM   FIXED SHAPED BEAM   VARIABLE SHAPED BEAM   CHARACTER PROJECTION

1   25   150   450

NUMBER OF IMAGE POINTS ADDRESSED SIMULTANEOUSLY

Figure 43. Comparison of pattern generation techniques and resultant profiles.

2.2.b.2 Raster Scan Systems. A raster scan system employing a continuously moving stage, was described by Alles and Herriott in 1975 (*64*). This system, known as EBES, was developed to provide fast, practical, and economical exposure of master masks and special purpose devices. The design criteria were 0.50- and 0.25- μm round beam address structure, 1-μm minimum features using small-field, telecentric beam deflection. The continuously moving air bearing stage is laser controlled interferometrically in real time with an absolute accuracy of 0.125 μm over a 100- cm^2 pattern area. The chip pattern is broken down (as shown in Figure 40) into parallel stripes, with the pattern in one stripe being decoded with a small computer during stage retrace and stored in memory for repeated writing. Raster scanning is convenient, since it uses almost constant table velocity and uses information read from a memory buffer in a serial fashion, thereby minimizing the frequency response or bandwidth required in the deflection system. This system has demonstrated reliable and stable performance for both mask fabrication and direct wafer writing. Detailed descriptions of this machine and its performance can be found in many references in the literature.

Several other successful machines have also been reported using raster scanning. The IBM EL-1 is a step-and-repeat type system that exposes a 5 mm x 5 mm field. These small fields are then stitched together to form the complete circuit. The EL-1 systems use a 2.5 μm x 2.5 μm square beam and are used to manufacture custom circuits.

2.2.b.3 Projection systems. High throughput machines can be designed to project an entire circuit or wafer simultaneously. A one-to-one whole wafer projection system has been described by several laboratories (*67,68*). These systems rely on photoelectron emission from a UV-illuminated cathode (mask), and image those electrons using a large magnetic lens. Reduction electron projection systems have also been reported using either photocathode emitters or complex metal masks (*69*) (see Figure 44).

2.3 X-ray Lithography

X-rays with wavelengths in the 0.04- to 0.5- nm region represent another alternative radiation source with potential for high resolution pattern replication into polymeric resist materials. X-ray lithography is, in effect, a method used to obtain a one-to-one x-ray shadowgraph of a mask and recording the mask image in the underlying resist. In 1972, Spears and Smith demonstrated that one could obtain high resolution patterns using x-ray proximity printing and suggested that this technique could be used for high throughput, high resolution semi-conductor device fabrication (*70*). The essential ingredients in x-ray lithography include:

1. A mask consisting of a pattern made with an x-ray absorbing material on a thin x-ray transparent membrane,

2. An x-ray source of sufficient brightness in the wavelength range of interest to expose the resist through the mask, and

3. An x-ray sensitive resist material.

Figure 45 illustrates the various x-ray lithographic options that can be considered. Since there are no efficient x-ray imaging optics currently available, one is committed to designing lithography systems that are one-to-one image printers. As with photolithography, one-to-one full wafer printers can be designed or, alternatively a step-and-repeat strategy can be employed to eliminate some problems in registration and mask stability. Currently there are two strong contenders for x-ray radiation sources: 1) electron impact and 2) synchrotron sources. Conventional electron impact sources produce a broad spectrum of x-rays, centered about a characteristic line of the material, which are generated by bombardment of a suitable target material by a high energy electron beam. The synchrotron or storage ring produces a broad spectrum of radiation stemming from energy loss of electrons in

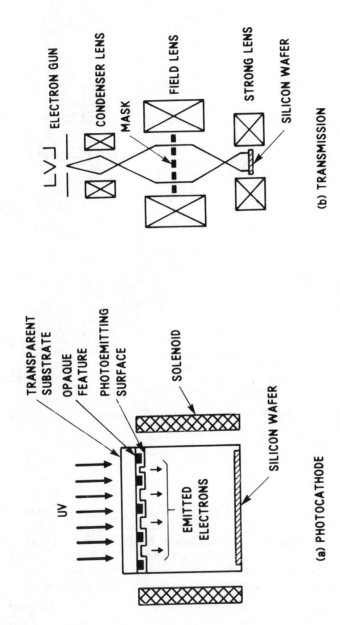

Figure 44. Schematic of electron beam image projection systems: (a) image formed by electrons emitted from a photocathode, and (b) image formed by an absorbing mask pattern on an electron transparent substrate.

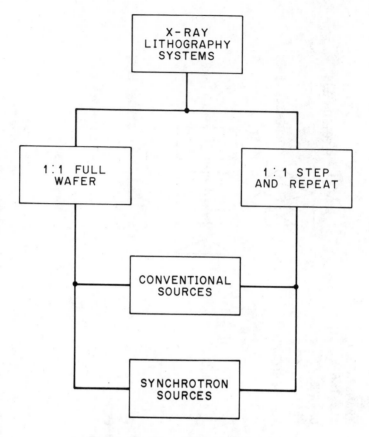

Figure 45. X-ray lithographic options

motion at relativistic energies. This radiation is characterized by an intense, continuous spectral distribution from the infrared to the long wavelength x-ray region. It is highly collimated and confined near the orbital plane of the circulating electrons, thereby requiring spreading in the vertical direction or moving the mask and wafer combination with constant speed through the fan of synchrotron radiation. Synchrotrons offer the advantage of high power output.

2.3.a X-ray Interactions. For all practical purposes, x-rays travel in a straight line with their scattering range equal to the photoelectron energy produced in the material following the absorption of an x-ray photon. The energy absorption can be calculated from the exponential dependence on the mass absorption coefficient, μ_m, the density, ρ and the thickness, z, of the absorbing film according to

$$I = I_o \, e^{-u_m \rho z} \tag{48}$$

where I_o and I are the intensities of the x-rays before and after passage through the absorbing film. Absorption of an x-ray photon results in the formation of a photoelectron which undergoes elastic and inelastic collisions within the absorbing material producing secondary electrons which are responsible for the chemical reactions in the resist film. The range of the primary photoelectrons is on the order of 100-200 nm. and does not represent a significant resolution limitation for this technology which is limited, from a practical standpoint, by several types of distortions that stem from the geometry of the exposure system and distortions and inaccuracies in the mask. A major limitation is that of penumbral shadowing depicted in Figure 46. Since the x-ray source is finite in size and separated from the mask by a distance D, the edge of the mask does not cast a sharp shadow

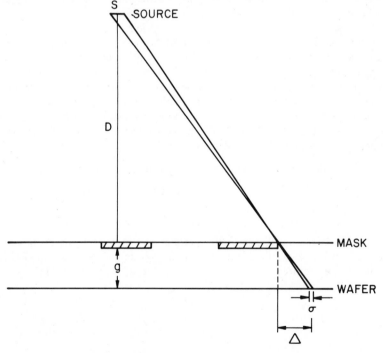

Figure 46. Geometry involved in x-ray exposure.

but has a region associated with it which is known as a *penumbral blur, σ*. This blur at the edge of the feature can be calculated from the geometry of Figure 46 via Equation 49 which relates σ to the source size d_s, the separation between the source and the mask and the separation between the mask and wafer g.

$$\sigma = g \, \frac{d_s}{D} \tag{49}$$

In addition to penumbral blur, there is also a small magnification factor (Δ) at the edge of the mask. This is termed run-off distortion and results from the separation of source and mask as shown in the following equation:

$$\Delta = g \tan \theta \tag{50}$$

where θ is the position-dependent angle of incidence of the x-ray beam on the mask. The small magnification, which increases linearly from the center where it is zero to some finite value (several microns) at the edge of the pattern, is of little concern for multi-layer devices, since it will have the same value for each level, providing all parameters remain constant within a given exposure tool.

An additional factor that degrades the resolution is mask contrast. Figure 47 shows the linear absorption coefficient, α, of several materials which may be used as absorbers and substrates for x-ray masks. Except for discontinuities at absorption edges, absorption increases with increasing wavelength. It is interesting to note that the most absorbing materials are only about a factor of 50 higher in their absorption coefficient than the most transparent materials. This implies that mask substrates must be relatively thin in order to have sufficient transmission, and that the absorber pattern must be relatively thick to attenuate the x-rays sufficiently. Mask contrast is the ratio of the transmission in the clear and opaque regions of the mask. A low contrast mask will result in exposure of the resist in nominally opaque regions. Additional exposure is caused by:

1. Continuous x-rays that are not absorbed by the mask,

2. Secondary and Auger electrons ejected from the mask surface, and

3. Secondary electrons ejected from the resist substrate.

The x-ray mask represents perhaps the most difficult component to fabricate in the x-ray lithographic system. Watts and Maldonado (71) have outlined the requirements for a mask capable of replicating VLSI patterns in a production environment as follows:

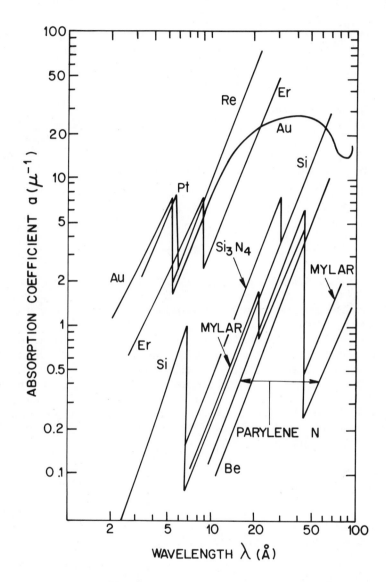

Figure 47. Absorption coefficient of some of the most absorbing and most transparent materials for soft x-rays.

1. The substrate must be transparent to the source radiation in the range of maximum resist sensitivity;

2. The mask absorber must be capable of providing sufficient attenuation to the source radiation in the range of maximum resist sensitivity;

3. The mask must be rugged and capable of being handled by production workers;

4. The substrate must be dimensionally stable, i.e., The thermal expansion properties of the mask substrate must match those of the wafer and mask support structure; Also all of the positional and dimensional changes of the mask features during processing should be smaller than the maximum allowable registration error;

5. Photoelectrons generated on the mask should be prevented from reaching the resist surface;

6. The defect density should be low (less than one defect per sq. cm.);

7. The mask must be transparent to the radiation used for alignment.

Figure 48 represents one type of mask structure that has been developed at Bell Laboratories and used successfully to replicate devices with one-micron features. The reader is referred elsewhere for a detailed discussion of mask fabrication procedures (72).

Figure 49 is a simplified diagram of an x-ray proximity printer showing the source, x-ray confinement cavity, mask, and wafer. As discussed previously, it is necessary to design a source with as small a radiating area as possible in order to minimize penumbral shadowing and with a maximum intensity of x-rays to minimize exposure time. Watts and Maldonado have extensively reviewed conventional x-ray sources, and the reader is referred to that work for additional details.

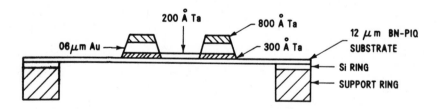

Figure 48. Typical x-ray mask.

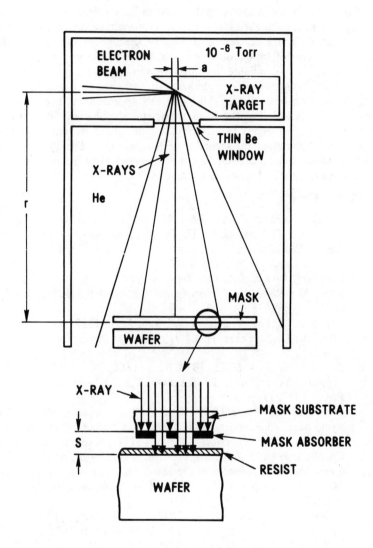

Figure 49. Schematic of an x-ray exposure system.

2.4 Ion Beam Lithography

It has been shown recently that when an accelerated ion passes through matter it can induce chemical reactions in a manner similar to those caused by electrons. Since scattering of ions in the one to three MeV range is several orders of magnitude less than that for electrons, Ion beam lithography is a candidate for high resolution lithography (*73,74*). The focused ion beams can be electrostatically scanned in a manner similar to that in electron beam lithography, and this approach has been used to generate high resolution patterns in a variety of polymeric resist materials. Resolution in ion beam lithography is determined by ion scattering and scattering of the resultant secondary electrons. This technique is being studied in research laboratories, and much work is required before ion beams can be contemplated for device manufacturing.

Literature Cited

1. Bowden, M. J. *J. Electrochem. Soc.* 1981, *128* (5), p. 195C.
2. Widmann, D.; Stein, K. U. in "Semiconductor Technologies with Reduced Diminsions, Solid State Circuits", Proc. 2nd Eur. Solid State Circuits Conf. 1976, *29*, p. 1977.
3. Cuthbert, J. D. *Solid State Technol.*, 1977, *20*, p. 59.
4. Born, M.; Wolf, E. "Principles of Optics" 4th Ed. Pergamon Press, N.Y. 1970; p 559.
5. Lin, B. J. *J. Vac. Sci. Technol.* 1975, *12*, p. 1317.
6. McGillis, D. A.; Fehrs, D. L. *IEEE Trans. Elect. Dev.* 1975, *ED-22*, p. 471.
7. Smith, H. I. *Proc IEEE,* 1974, *62*, p. 1361.
8. Mimura, Y.; Ohkubo, T.; Takeuchi, T.; Sekikawa, K. *Jap. J. Appl. Phys.* 1978, *17*, p. 541.
9. Doane, D. A.; Fraser, D. B.; Hess, D. W. "Proceedings of the Tutorial Symposium on Semiconductor Technology," 1982, *82* (5), p. 119.
10. Markle, D. A. *Solid State Technol.* 1974, *50* (6), p. 68.
11. Cuthbert, J. D. *Solid State Technol.* 1977, 59(6), p. 42.
12. King, M. C. "Principles of Optical Lithography" in *VLSI Electronics Microstructure Science*, 1981, *1*, ed. by Einspruch, N. G., Academic Press, New York, N.Y., p 41-81.
13. Wilcznski, J. S.; Tibbetts, R. *IBM J. Res. Develop.* 1969, *13*, p. 192.
14. Bruning, J. H. *J. Vac. Sci. Technol.* 1980, *17*, p. 1147.
15. Tobey, W. *Electronics* 1977, *50*, p. 115.
16. Wittekiek, S. *Electronics* 1977, *50* (10), p 32.
17. Lacombat, M. *Proc. Int. Conf. on Microlithography* 1977, p 83.
18. Wilczynski, J. S.; Tibbetts, R. *IBM J. Res. and Develop.* 1969, *13*, p 192.

19. Roussel, J. *Solid State Technol.* 1978, 5, p 67.
20. Wittekoek, S. *Solid State Technol.* 1980, 6, p 80.
21. O'Toole, M. M.; Neureuther, A. R., *Develop. in Semicond. Microl. SPIE.* 1979, *174*, p 22.
22. Hershel, R. *J of SPIE*, 1978, *135*, p 48.
23. Hopkins, H. H., *J.O.S.A.* 1957, *47*, p 508.
24. Swing, R. E.; Clay, J. R. *J. Opt. Soc. Am.* 1967, *57*, p. 1180.
25. Born, M.; Wolf, E. "Principles of Optics; Pergamon Press: New York, N.Y., 1970; p 550.
26. Blanchard, C. H.; Burnett, C. R.; Stoner, R. G.; Weber, R. L. "Introduction to Modern Physics"; Prentice-Hall, Inc; Englewood Cliffs, N.J., 1963, Chap. 5.
27. Goodman, J. W. "Introduction to Fourier Optics"; McGraw Hill: New York, 1968, Chap. 1-3.
28. Lacombat, M.; Massin, J.; Dubroeuck, G. M.; Brevignon, M. *Solid State Technol.* 1980, 23, p 115.
29. Levi, L.; Austing, R. H. *Appl. opt.* 1968, 7, p 967.
30. Bruning, J. H. *J. Vac. Sci. Technol.* 1979, 16, p 1925.
31. Dill, F. H. *IEEE Trans. Electron Devices* 1975, *ED-22*, p 440.
32. Dill, F. H.; Neureuther, A. R.; Tutt, J. A.; Walker, E. J., 1975, *ibid*, *ED-22*, p 456.
33. Walker, E. J. *IEEE Trans. Electron Devices* 1975, *ed-22*, 456.
34. Brewer, G. R. "Electron-Beam Technology in Microelectronic Fabrication"; Academic Press: New York, 1980, Chap. 1-4.
35. Ahmed, H.; Nixon, W. C. "Microcircuit Engineering"; Cambridge Univ. Press: Cambridge England, 1980, pp 1-180.
36. Mollenstedt, G. *Third Symp. Electron Beam Technol. Proc.* 1961, pp 340-359.
37. Oatley, C. W. *J. Phys. E.* 1975, 8, p 337.
38. Chang, T. H. P.; Nixon, W. C. *J. Sci. Instrum.* 1967, *44*, p 230.
39. Hawryluk, R. J., *J. Vac. Sci. Technol.*, 1981 19(1), p 1.
40. Nosker, R. W., *J. Appl. Phys.*, (1969) p 40, 1872.
41. Hawryluk, A. M.; Hawryluk, R. J.; Smith, H. I. *J. Appl. Phys.* 1974, *45*, p 2551.
42. Kyser, D.; Viswanathan, N. S. *J. Vac. Sci. Technol.*, 1975, *12*, p 1305.
43. Greeneich, J. S.; Van Duzer, T. *IEEE* Trans. Electron, Devices 1974, *ED-21*, p 286.
44. Heidenreich, R. D. *J. Appl. Phys.* 1977, *48*, p 1418.
45. Everhart, T. E., *J. Appl. Phys.* 1960, *31*, p 1483.
46. Chung, M. S. C., Tai, K. L. Proc. 8th Intl. Conf. on Electron and Ion Beam Sci and Technol., R. Bakish, Ed., 1978, p 242, Electrochemical Soc.

47. Heidenreich, R. D.; Thompson, L. F.; Feit, E. D.; Melliar-Smith, C. M. *J. Appl. Phys.* 1973, *44*, p 4039.

48. Heidenreich, R. D.; Ballantyne, J. P.; Thompson, L. F. *J. Vac. Sci. Technol.* 1975, *12*, p 1284.

49. Everhart, T. E.; Hoff, P. H. *J. Appl. Phys.* 1971, 42, p 5837.

50. Ku, H. Y.; Scala, L. C. *J. Electrochem. Soc.* 1969, *116*, p 980.

51. Greeneich, J. S. *J. Electrochem. Soc.* 1975, *122*, p 970, *J. Appl. Phys.* p 1974, *45*, p 5264.

52. Wolf, E. D.; Ozdemir, F. S.; Perkins, W. E.; Coane, P. J., Proc. 11th Symp. on Electron, Ion and Laser Beam Technol., R. F. Thomley Ed., San Francisco Press, 1971 p. 331.

53. Hatzakis, M. *J. Vac. Sci. Technol.* 1975, *12*, p 1276.

54. Hatzakis, M.; Ting, C. H.; Viswanathan, N., *Proc. 6th Intl. Conf. on Electron and Ion Beam Sci. and Technol.,* R. Bakish, Ed., Electrochemical Soc., 1974, p 592.

55. Chang, T. H. P.; Nixon, W. C. *J. Sci. Instrum.* 1967, *44*, p. 230.

56. Chang, T. H. P.; Wilson, A. D.; Speth, A. J.; Ting, C. H. *Procd. 7th conf. on Electron and Ion Beam Science and Technol* 1976, b, p. 392.

57. Ballantyne, J. P. *J. Vac. Sci. Technol.* 1975, *12*, p. 1257.

58. Herriott, D. I. "Proc. Tutorial Symp. on Semiconductor Technol." Electrochemical Soc., 1982, *82-5*, p. 139.

59. Herriott, D. R.; Brewer, G. R. "Electron-Beam Lithography Machines" in *Electron-Beam Technology in Microelectronic Circuit Fabrication*; Academic Press, New York, NY, 1981; pp. 141-216.

60. Burn, M.; Wolf, E. "Principles of Optics" 4th Ed. Pergamon Press, N.Y. 1970; p. 559.

61. Grivet, P. "Electron Optics", Pergamon Press, Oxford, Eng., 1965.

62. Pfeiffer, H. C. *J. Vac. Sci. Technol.* 1978, *15*, p. 887.

63. Varnell, G. L.; Spicer, D. F.; Habley, J. F.; Robbins, R. R.; Carpenter, C.; Malone, J. J. *J. Vac. Sci. Technol.* 1980, *16*, p. 1787.

64. Herriott, D. R.; Collier, R. J.; Alles, D. S.; Stafford, J. W. *IEEE Trans. Electron Devices* 1975, *ED-22*, p. 385.

65. Alles, D. S.; Ashley, F. R.; Johnson, A. M.; Townsend, R. L. *J. Vac. Sci. Technol* 1975, *12*, p. 1252.

66. Yourke, H. S.; Weber, E. V. *Tech. Digest IEDM*, 1976, Washington pp. 431-436.

67. Scott, J. P. *J. Appl. Phys.* 1975, *46*, p. 661.

68. O'Keefe, T. W.; Vine, J.; Handy, R. M. *Solid State Electronics*, 1969, *12*, p. 841.

69. Herritage, M. B. *J. Vac. Sci. Technol.* 1975, *12*, p. 1135.

70. Spears, D. L.; Smith, H. I. *Solid State Technol.* 1972, *15*, p. 21.

71.   Watts, R. K.; Maldonado, J. R. "VLSI Electronics: Microstructure Science, Vol. 4" Academic Press, New York, NY, 1982, pp. 56-97.
72.   Maydan, D.; Coquin, G. A.; Levenstein, H. J.; Sinha, A. K.; Wang, D. N. K. *J. Vac. Sci. Technol.* 1979, *16*, p 1959.
73.   Hall, T. M.; Wagner, A.; Thompson, L. F. *J. Vac. Sci. Technol.* 1979, *16*, p 1889.
74.   Seliger, R. L.; Kubena, R. L.; Olney, R. D.; Ward, J. W.; Wang, V. *J. Vac. Sci. Technol.* 1979, *16*, p. 1610.

RECEIVED February 4, 1983

Organic Resist Materials — Theory and Chemistry

C. Grant Willson

IBM Research Laboratory, San Jose, CA 95193

0097-6156/83/0219-0087$16.95/0

3.1 Introduction

The word lithography is defined by Webster as "the art or process of putting writings or designs on stone with a greasy material and producing printed impressions therefrom; also, any process based on the same principle, as one using zinc, aluminum, or some other substance instead of stone." Webster's definition is a reasonable description of the art as it is practiced in the semi-conductor industry even though the definition was written many years before the advent of the semiconductor device. We still write on stone, if one considers single crystal silicon a special case of stone, and the greasy material described in the Webster definition is, in fact, the material we now know as resist.

The basic steps of the lithographic process are outlined in Figure 1. The example shown corresponds to photolithography in which the resist material is applied as a thin coating over some base and subsequently exposed in an image-wise fashion (through a mask) such that light strikes selected areas of the resist material. The exposed resist is then subjected to a development step. Depending upon the chemical nature of the resist material, the exposed areas may be rendered more soluble in some developing solvent than the unexposed areas, thereby producing a *positive tone* image of the mask. Conversely, the exposed areas may be rendered less soluble producing a *negative tone* image of the mask. The net effect of this process is to produce a three dimensional relief image in the resist material that is a replication of the opaque and transparent areas on the mask.

The areas of resist that remain following the imaging and developing processes are used to mask the underlying substrate for subsequent etching or other image transfer steps. If, for example, the underlying substrate or base were SiO_2, immersion of the structure into an etchant such as buffered hydrofluoric acid would result in selective etching of the SiO_2 in those areas that were bared during the development step. The resist material "resists" the etchant and prevents it from attacking the underlying substrate in those areas where it remains in place after development. The generic name *resist* presumably evolved as a description of the ability of these materials to "resist" etchants (see Chapter 1, Figures 2 and 4 and Chapter 4). Following the etching process, the resist is removed by stripping, as indicated in Figure 1, to produce a positive or negative tone relief image in the underlying substrate.

Webster's definition includes the production of printed images from the designs one generates on "stone," and refers to any of three types of processes used in the printing industry. These include: 1) lithographic (or planographic) method, 2) intaglio method, and 3) typographic or relief printing. In the former process, a lithographic plate is imaged as described in Figure 1 and coated with ink. The ink is such that it selectively wets the

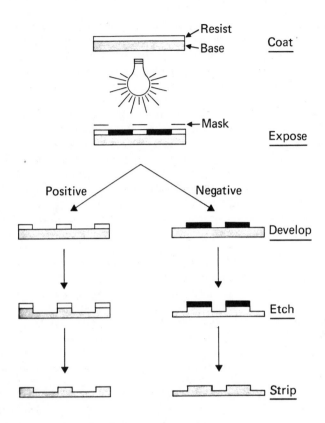

Figure 1. Schematic representation of the photolithographic process sequence.

resist areas on the lithographic plate. Pressing this ink-covered plate against paper transfers the resist image onto the paper as an inked pattern. In the intaglio method the lithographic plate is subjected to an etching step such that relief images are formed in the substrate or base layer. The resist is then stripped (removed), and the plate is coated with ink. Ink is retained in the *recessed* areas. When the inked plate is pressed against paper, the ink is transferred to the paper to produce an ink pattern of the recessed areas. The typographic process resembles, in many respects, the intaglio method with the exception that the ink is retained on the *raised* surfaces of the relief structure that is produced in the plate (*1*). The major commercial application of resist materials has been and continues to be in these printing processes where photolithography is used to generate master printing plates from which many copies of newspapers, books, etc. are subsequently generated. The economics of the printing industry demand that transfer of the image from the master plate to the paper must be rapid, but since one plate can produce thousands and thousands of copies, the *primary* resist imaging process used to make the master plate need not be fast.

The resist technology as practiced by the semi-conductor industry was borrowed from the printing industry; yet the demands that semi-conductor manufacturing places on resist materials are extreme when compared to those required for production of master printing plates. Manufacturing of semi-conductor devices requires the generation of several sequential *primary* images that define the diffusion, insulator, and conductor lines that make up devices. Here, unlike the production of printing plates, the sensitivity of the resist material becomes a major issue, since the final product is the resist image itself and not a paper or book printed from that image.

The resist materials that were developed for the printing industry are also useful for the manufacture of semi-conductor devices. In retrospect, this is fortuitous since the demands of the two technologies are quite different. Consider, for example, the resolution required to print a newspaper where the goal is to generate legible print. Here the resolution need only be a fraction of a millimeter, whereas typical semi-conductor devices that are in production today have minimum features of two or three microns. These features are smaller than a typical bacterium and are comparable in size to the organelles of cells.

In order for a resist to be useful in semi-conductor manufacturing, it not only must have high sensitivity and the ability to resolve such small features, but also must be capable of being spincoated into thin and continuous films that will adhere to a variety of substrates ranging from metals and semi-conductors to insulators. It should also be able to withstand exposure to extremely high temperature and exceedingly corrosive etching environments such as strong acids and plasmas, without loss of adhesion or line definition. The purpose of this chapter is to familiarize the reader with some of the chemical principles involved in the design and development of these remarkable materials we call resists.

3.2 The Resist Hierarchy

Resist materials can be classified as positive or negative on the basis of their radiation response as described in Section 3.1 and illustrated in Figure 1. Both resist types can be subdivided into two categories depending upon the basic nature of their design: 1) one-component systems and 2) two-component systems (see Figure 2). One-component systems are polymers that combine radiochemical reactivity with etch resistance and film-forming characteristics. In two component systems, the resist is formulated from an

Resist

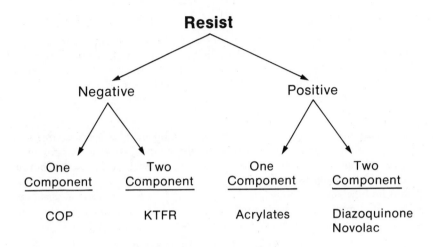

Figure 2.   The resist hierarchy.

inert matrix resin (which serves only as a binder and film-forming material) and a sensitizer molecule which, in general, is monomeric in nature and undergoes the radiochemical transformations that are responsible for imaging.

The familiar positive photoresists, Hunt's HPR, Shipley's Microposit, Azoplate's AZ etc., are all *two-component,* resist systems, consisting of a phenolic resin matrix material and a diazonaphthoquinone sensitizer. The matrix material is essentially inert to photochemistry and was chosen for its film-forming, adhesion, chemical and thermal resistance characteristics. The chemistry of the resist action only occurs in the sensitizer molecule, the diazonaphthoquinone. A detailed description of these materials, their chemical structures and radiation chemistry will be discussed in Section 3.5.b.

Poly(methyl methacrylate) (PMMA) is a classical *one-component, positive* resist system. PMMA is a single, homogeneous material that combines the properties of excellent film-forming characteristics, resistance to chemical etchants and intrinsic radiation sensitivity.

The most familiar negative photoresists are examples of *two-component,* resist materials. These include Kodak's KTFR, Merck's Selectilux N, Hunt's HNR, etc., all of which consist of a cyclized synthetic rubber matrix resin which is radiation insensitive but forms excellent films. This resin is combined with a bis-arylazide sensitizer.

COP, the familiar negative e-beam resist developed at Bell Laboratories, is an example of a *one-component negative* resist system. COP is a copolymer which has excellent film-forming characteristics, resistance to etchants, and intrinsic radiation sensitivity.

3.3 Resist Sensitometry

There are two aspects of resist sensitometry: 1) the measurement of resist response to radiation which involves measurements designed to determine the intrinsic radiation sensitivity of the materials from which the resist is constituted and 2) lithographic sensitivity which is a measure of the efficiency with which these radiochemical conversions can be expressed in relief image formation.

3.3.a Photo-efficiency. First consider two-component resist systems. In these formulations the photosensitivity is determined, to a first approximation, by the photo-efficiency of the sensitizer molecules. Photoefficiency is quantitatively defined in terms of the photochemical quantum efficiency. Quantum efficiency, ϕ, is defined by Equation 1 as the number of sensitizer molecules converted to photo-product divided by the number of absorbed photons required to accomplish that conversion.

$$\phi = \frac{\text{\# of molecules transformed}}{\text{\# of photons absorbed}} \tag{1}$$

Experimental methods for determining ϕ are well documented (*2*). These experiments are conveniently carried out and require only a method of producing reasonably narrow-bandwidth radiation, a method of measuring the flux of that radiation per unit area, and a UV-visible spectrophotometer. The quantum efficiency of typical diazonaphthoquinone sensitizers of the type that are used in the formulation of positive photoresists ranges from 0.2 to 0.3, whereas the quantum efficiency of the bis-arylazide sensitizers used in the formulation of two-component negative photoresists, ranges from 0.5 to 1.0.

It is instructive to consider the definition of photochemical quantum efficiency and its relationship to lithographic sensitivity. In order to achieve maximum lithographic sensitivity, the resist designer hopes to achieve the maximum number of moles of sensitizer converted to photo product, per unit time, for a given radiation flux incident at the resist surface. This value is directly proportional to ϕ, the quantum efficiency, and to the number of photons absorbed. Therefore, it is possible to increase resist sensitivity both by increasing the quantum efficiency through judicious choice of the sensitizer structure and, by increasing the number of photons absorbed per unit time at constant flux, by modifying that structure for maximal absorbance at the wavelength of interest.

Experimental determination of the quantum efficiency of photosensitive polymers of the sort that are used in one-component positive resist systems is a more complex experimental undertaking. Here the quantum efficiency is defined as the number of main chain scissions that occurs per photon absorbed. Guillet and coworkers at the University of Toronto have designed and built an apparatus which allows convenient determination of the quantum efficiency of scission for polymers in solution (*3,4*). They have applied this measurement technique to a range of poly(vinyl ketones) and poly(vinyl ketone) copolymers in solution. Their measurements indicate that a typical vinyl ketone polymer such as poly(methyl isopropenyl ketone) (PMIPK) has a solution quantum efficiency of approximately 0.3 at 313 nm. Guillet's studies also clearly indicate that the quantum efficiency of vinyl ketone polymers is substantially higher in solution than in the glassy state. When photopolymers are used as resist materials, they are always in the solid state, and hence care must be exercised in attempting to predict resist sensitivity from a comparison of polymer quantum efficiencies measured in solution. It is possible, though far more tedious, to determine the quantum efficiency of photopolymers in the solid state. Such data would be useful in directly predicting resist performance, but unfortunately there is only a limited amount available.

3.3.b Ionizing Radiation Efficiency. Experimental methods for the determination of the radiation sensitivity of polymers that undergo net scission upon exposure to high energy e-beam, x-ray or gamma radiation, are well documented. PMMA is a convenient material for such studies. Figure 3 shows the gel permeation chromatograms of a series of PMMA samples that have been subjected to increasing doses of cobalt 60 gamma radiation. Note that the molecular weight decreases monotonically with increasing dose. Quantitative analysis of the molecular weight response to dose in polymers that undergo net scission has been published by several authors (*5-7*) and leads to an important relationship between molecular weight and dose which can be derived simply.

Suppose we have a sample of polymer of weight w grams containing N_o molecules. By definition, the number average molecular weight M_n^o is given by

$$M_n^o \equiv \frac{wN_A}{N_o} \qquad (2)$$

where N_A is Avogadro's number. Rearranging Equation 2 yields

$$N_o = \frac{wN_A}{M_n^o} \qquad (3)$$

for the total number of molecules in the sample prior to radiolysis. The sample is now irradiated to a dose, D, in eV/g. The total dose absorbed by the sample is Dw eV and since the *total number of scissions* produced in the sample, N^*, is proportional to the absorbed dose, we may write

$$N^* = KDw \qquad (4a)$$

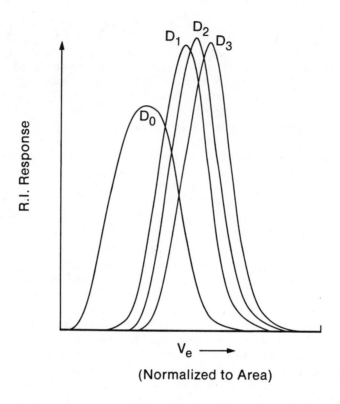

Figure 3.   Gel permeation chromatograms of PMMA exposed to increasing doses of γ-radiation.   High molecular weight fractions elute at low elution volumes.

where K is a structure-dependent constant. This constant is generally expressed in terms of a G-value, $G(s)$, where $G(s)$ is the number of chain scissions produced per 100 eV of absorbed energy. Thus Equation 4a can be re-written as

$$N^* = (G(s)/100)Dw \qquad (4b)$$

Since each scission event increases the number of molecules by one, the number average molecular weight after exposure to dose D will be given by

$$M_n^* = \frac{wN_A}{N_o + N^*} \qquad (5)$$

where the total mass of the polymer is assumed to remain constant. $N_o + N^*$ is the total number of molecules in the sample after exposure. Substituting Equations 3 and 4b into Equation 5 yields

$$M_n^* = \frac{wN_A}{wN_A/M_n^o + (G(s)/100)Dw} \qquad (6)$$

$$= \frac{N_A}{N_A/M_n^o + (G(s)/100)D} \qquad (7)$$

which is independent of the sample mass. Rearranging Equation 7 into the form of Equation 8, i.e.,

$$\frac{1}{M_n^*} = \frac{1}{M_n^o} + (G(s)/100N_A)D \qquad (8)$$

allows one to see immediately that a linear relationship exists between inverse number average molecular weight after irradiation and dose. Thus a plot of $1/M_n$ vs. dose should produce a straight line with intercept $1/M_n^o$ and a slope which is directly proportional to $G(s)$. For a rigorous derivation, the reader is referred to reference (8).

As we mentioned before, $G(s)$ is a structure dependent constant that describes the number of scissions per unit absorbed dose and in that sense can be compared to a photochemical quantum efficiency. The $G(s)$ of a radiation-sensitive polymer is a figure of merit that can be used in comparing one material with another. There is a very high correlation between $G(s)$ values to gamma radiation (the radiation most commonly used for determining $G(s)$) and high sensitivity for lithographic materials used in either electron beam, ion beam or x-ray exposure.

Figure 4 is a plot of inverse number average molecular weight vs. gamma radiation dose for three polymers: nitrocellulose, PMMA, and poly(α-hydroxyisobutyric acid). The purpose of this experiment was to determine whether nitrocellulose or poly(α-hydroxyisobutyric acid) might be a more sensitive electron beam resist material than PMMA (9). An examination of the plot indicates that poly(α-hydroxyisobutyric acid) undergoes far fewer scissions per unit dose than PMMA and is, therefore, unlikely to be a more sensitive resist. The slopes of the lines ($G(s)$ values) for PMMA and nitrocellulose are virtually identical. The analysis, therefore, predicts that the potential sensitivity of nitrocellulose as an e-beam resist material is unlikely to be significantly greater than that of PMMA.

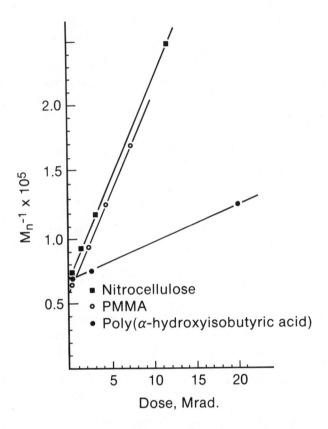

Figure 4. Plot of inverse number average molecular weight vs dose for three polymers. The slope of the lines is proportional to $G(s)$.

The use of gamma radiation in the determination of scission G-values offers great experimental simplicity. Because the absorption coefficients of organic polymers are low for gamma radiation, the absorbed dose in bulk samples can be considered constant with thickness. Consequently, one needs only to seal samples of the polymer into glass ampules, expose the ampules to a known dose of gamma radiation, and then measure the molecular weight of the irradiated samples either by membrane osmometry or gel permeation chromatography.

To conduct such an experiment using electron beam radiation is more tedious, since the absorption cross section of organic materials to electrons is higher than that for gamma rays. As a consequence, the samples must be exposed as thin films of known thickness, then dissolved and subjected to molecular weight determination. This experiment has been conducted in at least one case (*10*), the data from which is presented in Figure 5. Note that a plot of inverse number average molecular weight vs. dose of electron beam radiation in microcoulombs per cm^2 ($\mu C/cm^2$) provides a line similar to that depicted in Figure 4. If Figure 5 is replotted in terms of energy *deposited* per unit volume rather than *incident* dose, a G (scission) value is obtained that is remarkably consistent with that from γ-ray experiments.

So far we have only considered polymers that undergo main-chain scission upon exposure to radiation. PMMA is an example of such a material. If, on the other hand, one considers polymeric systems in which both scissioning and crosslinking events occur simultaneously upon exposure, the analysis depicted above will allow determination only of the *net* scission-

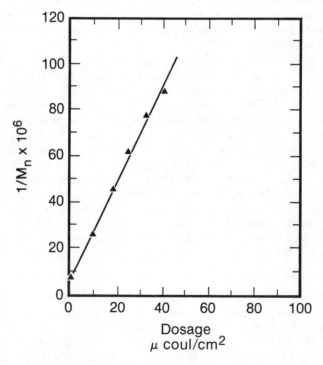

Figure 5.   Inverse number average molecular weight vs dose of PMMA exposed to electron beam radiation.

ing efficiency. If only scissioning events occur upon exposure as in the case of PMMA, the dispersity of the polymer as characterized by M_w/M_n will approach a value of 2 with increasing dose. If, on the other hand, the material of interest undergoes both scissioning and crosslinking events upon exposure, the dispersity will increase with increasing dose.

It is possible to uniquely determine both the scission efficiency $G(s)$ and the crosslinking efficiency $G(x)$ via Equations 9 and 10 providing both the number average molecular weight (M_n) and the weight average molecular weight (M_w) can be measured.

$$\frac{1}{M_n^*} = \frac{1}{M_n^o} + [G(s) - G(x)](D/100N_A) \qquad (9)$$

$$\frac{1}{M_w^x} = \frac{1}{M_w^o} + [G(s) - 4G(s)](D/200N_A) \qquad (10)$$

Equations (9) and (10)) are simultaneous equations which relate initial and final number average molecular (M_n) weight to the variables $G(x)$ and $G(s)$ and the initial and final weight average molecular weight (M_w) to the same variables (11-12). Weight average molecular weight is most conveniently determined by low-angle, laser light scattering experiments or by gel permeation chromatography.

The final case to consider is that in which $G(x)$ is greater than $G(s)$. In such cases the polymer undergoes crosslinking (network formation) upon exposure as occurs with one-component, negative resists. Quantification of $G(x)$ in polymers that undergo net crosslinking can be carried out by exposing samples of the material to a range of gamma radiation doses and then exhaustively extracting the gel fraction with an appropriate solvent. A plot of the mass fraction of insoluble, crosslinked material vs. dose enables $G(x)$ to be determined (5). The $G(s)$ of polymers commonly used as one-component positive resist systems, expressed as the number of scissions per 100 eV of absorbed dose, ranges from 1.3 for PMMA to approximately 10 for certain poly(olefin sulfones). The $G(x)$ of common organic materials, expressed as number of crosslinks per 100 eV absorbed dose, ranges from 0.1 for poly(ethylene) to approximately 10 for polymers that contain oxirane groups (epoxy groups) in their side chains.

3.4 Lithographic Sensitivity

The measurement of intrinsic radiation sensitivity of various materials (defined by ϕ, $G(s)$ or $G(x)$), in one laboratory correlates well with measurements made in other laboratories. Measurement of lithographic sensitivity, on the other hand, is not nearly as precise. The literature is pervaded by papers describing resist sensitivity simply in terms of dose per unit area without the relevant experimental details. Interpretation of such results and their utility in comparing one resist with another demands extreme caution.

Resist sensitivity expressed in terms of dose per unit area is like an EPA mileage rating and at best should be used for comparison only.

Frequently the relationship between resist sensitivity reported in the literature and the dose that is required to successfully process that resist in a manufacturing environment is a tenuous one at best (see Section 4.2.a). Attempts to establish a standard method for determining resist sensitivity have been unsuccessful to date but are continuing. The difficulty with making such measurements and interpreting their meaning, stems from the fact that the production of lithographically useful relief images is dependent not only on the basic chemical nature of the resist material under study, but also upon the nature of the radiation used and a variety of other parameters such as, 1) the accelerating potential and dose rate of e-beam exposures, 2) the dose rate, bandwidth and wavelength of x-ray radiation and 3) the bandwidth of optical exposures. Processing parameters also affect sensitivity. These include the baking conditions, resist thickness, developer composition, development conditions, etc.

Measuring the sensitivity of positive-tone resist systems is more difficult than measuring the sensitivity of negative-tone systems and optical resists are more difficult to study than e-beam or x-ray resists. The following sections elaborate experimental methods that are routinely used in our laboratory to compare one resist with another. We have found that the sensitivity, defined and measured as described, provides us with both a figure-of-merit that has predictive utility and certain material parameters that are useful in computer process simulation studies that are beyond the scope of this book (*14,15*) (see Section 2.1). Other methods of defining resist sensitivity are discussed in Chapters 2 and 4. The choice of method must depend upon the goals of the study and will, in all cases, require careful interpretation.

3.4.a Positive Resist Sensitivity. Data of the sort presented in Figure 6 provides a great deal of quantitative information about the sensitivity, or dose response, of a positive resist system. It depicts the dissolution kinetics of an experimental e-beam resist system as a function of dose. The vertical axis is thickness, the horizontal axis time in the developer. At time = O_α, a film initially 1 μm thick, is immersed in an appropriate developing solvent. The change in thickness with time as a function of dose is presented as a family of curves (Figure 6). The dissolution rate of the unexposed resist is relatively slow at first, then increases in rate with depth, until finally the resist is developed to the substrate in approximately 1400 sec. Exposure of the resist to 10 $\mu C/cm^2$ of electron beam radiation allows the irradiated area to be removed in approximately 440 sec. of development.

The data presented in Figure 6 was accumulated using an FTA/1 instrument initially described by Konnerth and Dill of IBM (*16*). The

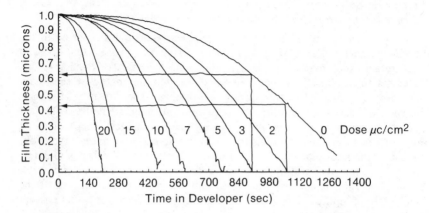

Figure 6. Dissolution kinetics as a function of dose for an experimental e-beam resist. Note that a 3 $\mu C/cm^2$ dose causes 1 μm of resist to dissolve in 848 sec at which time 0.62 μm of unexposed resist remains undeveloped. The data was generated on the FTAl1 Film Thickness Analyzer, Figure 7.

instrument utilizes thin film interferometric analysis to make rapid measurements of film thickness with great accuracy (Figure 7). Filtered light from a tungsten source is passed through a scanning monochrometer and reflected from the resist-substrate and resist-developer interfaces at the sample. The reflected light is collected on a photodiode. The diode output is digitized and then stored on an IBM Series 1 Computer. Analysis of the reflected intensity as a function of wave length and knowledge of the dielectric functions of the developer, resist and substrate allows calculation of the film thickness. The instrument can make one thickness measurement each microsecond.

A variety of other techniques have now evolved for generating this same sort of data. Noteworthy among these is the laser end-point detection system, diagrammed in Figure 8. This system uses a low-powered helium-neon laser directed through a bifurcated fiber optic cable onto the surface of a resist coated over a reflective substrate such as silicon. Light from the laser is reflected from both the resist-developer interface and from the resist-substrate interface. The two reflected rays are collected into the fiber optic cable and imaged onto a photodiode, the output of which is recorded on a strip chart. The reflected intensity, as a function of thickness, goes through a series of maxima and minima that represent constructive and destructive interference of the two reflected waves.

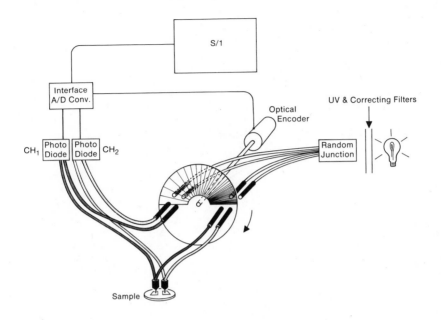

Figure 7.   FTA/1 Film Thickness Analyzer.

The output trace from the laser end-point detection system for a typical resist development cycle is depicted in Figure 9. The temporal distance between maxima can be related to a change in resist thickness through Equation 11

$$\Delta t = \lambda/2n \tag{11}$$

where λ is the wavelength of the laser, n is the index of refraction of the resist material at that wavelength and Δt is the change in thickness that occurs in the time span between maxima (or minima). All of the information required to produce the continuous curve depicted in Figure 6 is available in the trace shown in Figure 9. However, it is particularly easy from the data in Figure 9 to extract a curve such as that shown in Figure 10 by noting that the thickness at the end-point is 0 and each intensity maximum occurs at a time when the thickness is an increment Δt thicker than the succeeding one.

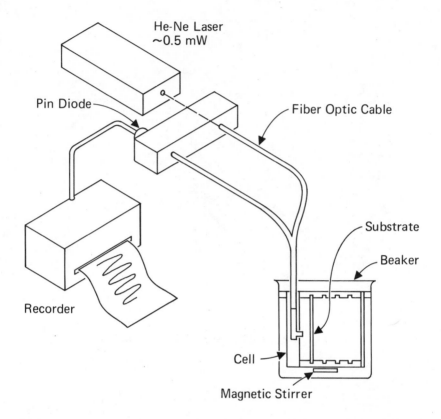

Figure 8. Laser End-point Detection System. Light from the laser is reflected off of the resist-substrate and resist developer interfaces. The reflected light is collected on a photodiode, the output of which is monitored on a strip chart recorder as a function of time. The output of this device is provided in Figure 9.

From the data available in plots such as Figure 6 or 10, a great deal of information regarding the lithographic sensitivity and performance of a resist can be determined. Figure 11 is derived from Figure 6 by plotting the thickness of unexposed resist remaining after development of an exposed area for that period of time required to just remove the last trace of resist from that exposed region vs. the logarithm of that dose. For example, when the resist in Figure 6 is exposed to $2\text{-}\mu C/cm^2$ of e-beam radiation, the

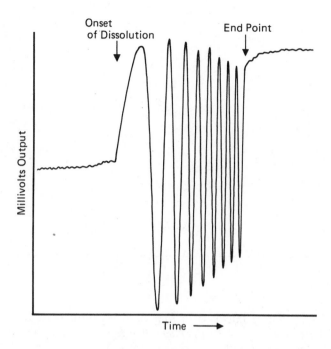

Figure 9.   Output of the Laser End-point Detection System.

exposed area reaches the substrate at a time when approximately 40% (4000 Å)of the initial thickness of the unexposed resist remains. For the 3-$\mu C/cm^2$ dose, the exposed area is clear when just over 60% (6000 Å) of the unexposed resist film remains. Extrapolation of the linear region of the plot in Figure 11 to the initial thickness (1-μm), provides a dose at which, in theory, 1-μm of the resist could be developed cleanly to the substrate without any significant loss of resist thickness in the unexposed region. We define this dose D_s, as the resist sensitivity. D_s is in qualitative agreement with D_p as defined in Section 4.2.a. For the experimental e-beam resist described in Figures 6 and 11, D_s is equal to approximately 12−$\mu C/cm^2$.

   When comparing one resist to another by this technique, it is necessary to hold some developing parameter constant. We have found that holding the unexposed development time constant by varying the developer strength provides data of predictive value. If therefore, another resist were to be compared with that described in Figure 6, the unexposed development time would be adjusted to approximately 1400 seconds and the curve

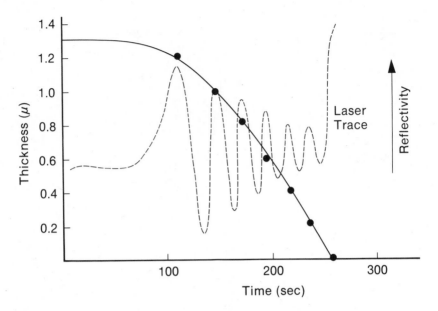

Figure 10.   A Plot of thickness vs time in developer derived from a laser end-point detector trace.   The thickness change between maxima is a function of the laser wavelength and the index of refraction of the resist at that wavelength.

corresponding to Figure 11 would be generated and the D_s values of the two materials compared. In all cases studied to date, the materials having the lower D_s in such a comparison have had the higher sensitivity in imaging experiments.

The relative contrast of two resists can also be compared by the analysis described above since resist contrast is related to the slope of the line in Figure 11. If the unexposed development time for two resists is made equal by developer formulation adjustments, the system with the steeper slope to the line used in extrapolation to D_s will have the higher contrast of the two. This comparison is useful **only** if the analyses are conducted under conditions of equal unexposed development time.

Experiments of the sort described above are easily carried out under conditions of e-beam or x-ray exposure. However, in the case of optical

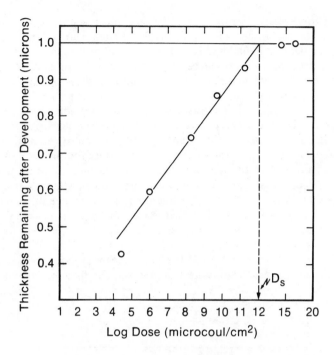

Figure 11. A sensitivity plot for a positive-tone experimental e-beam resist. The data is from Figure 8.

exposure, the same interference phenomena that produce the oscillating intensity function in Figure 9 are responsible for a substantial variation in total deposited energy with thickness and a variation in intensity with depth into the film (*17*). The latter variation is responsible for the familiar "standing wave" phenomenon. These interference-related energy variations complicate the determination of dissolution kinetics. Exposure to narrow-bandwidth radiation exaggerates the problem. The use of index-matched substrates or anti-reflection coatings under the resist can simplify the analysis but simultaneously reduce the signal to noise ratio in the thickness measurement. Because of these issues, extreme care must be exercised in the measurement of optical resist sensitivity and in interpretation of the resulting data.

3.4.b Negative Resist Sensitivity. Negative resist sensitivity is normally defined in terms of the incident dose of radiation required to produce a specified amount of insolublization. First consider a negative resist exposed to high energy radiation, such as an electron beam. In such a case, the measurement of resist sensitivity is a straightforward experiment. A nominal 1-μm thick film of the resist is spin-coated on an appropriate substrate, exposed to a range of doses, and developed in a suitable developer. The thickness of the insolubilized region remaining after development is then plotted against the log of the exposure dose, as indicated in Figure 12. Here, D_g^i (the gel dose) is the minimum dose required to produce an insolubilized residue. D_g^o is the minimum dose required to produce an insolubilized film of thickness equal to the initial film and γ, (the contrast of the resist) is defined as the slope of the line relating the change in insolubilized thickness with log of dose. The sensitivity of the resist is equal to D_g^x and represents the dose at which a lithographically useful image is formed (usually $\geqslant$50% gel formation).

Exposure of resist materials to electron beam radiation results in dose deposition throughout the thickness of the film for films of nominal 1-μm

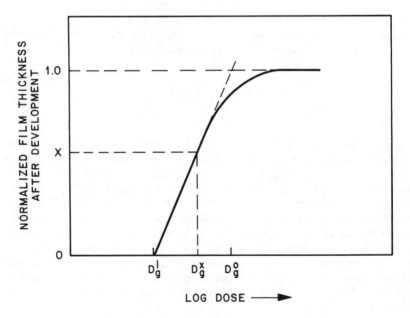

Figure 12.   A sensitivity plot for a negative e-beam resist.

thickness and accelerating potentials of greater than 5-10 kV. However, the highest dose per unit volume always occurs at the bottom of the film at the resist-substrate interface, where the accumulated dose is the sum of the incident dose (forward scattered) and the backscattered dose.

Conversely, optical exposure of negative resist films, in cases where the optical density of the film is substantial at the exposing wavelength, results in an energy deposition profile such that the *top* of the film receives a higher dose than the bottom of the film. Consequently, the first insolubilization of the resist film will occur at the surface rather than at the substrate, and in the case of underexposure, subsequent development will simply lift away the cross-linked material. Therefore, measurement of resist sensitivity in negative optical resists is best accomplished by casting these materials on a transparent substrate and exposing them *through* the substrate from the *back* side. In this fashion, the lowest exposure dose that will cause insolubilization will leave film attached to the substrate and allow generation of meaningful data.

The value D_s in units of incident dose per unit area for either a positive or negative resist system is of little value unless accompanied by a detailed description of the conditions under which it was measured. This description should include, at the minimum, the initial film thickness, the characteristics of the substrate, the temperature and time of the post- and pre-bake, the characteristics of the exposing radiation, and the developer composition, time and temperature. The structure, copolymer ratio, sequence distribution, molecular weight, and dispersity of polymers included in the formulation should also be provided.

3.5 The Chemistry of Classical Optical Resist Systems

3.5.a Two-Component Negative Resist. Until very recently, negative-tone optical resists have been the "workhorse" of the microelectronics industry. Billions of dollars worth of devices have been generated through use of these materials. Generically, they are bis-arylazide/rubber resists as disccused in Section 3.4 and are typified by Kodak's KTFR and MX118 Hunt's NMR, Merck's Selectilux-N, and a variety of similar formulations originating from U.S., European and Japanese manufacturers.

The matrix resin material used in the formulation of these resists is a synthetic rubber that is obtained by a Ziegler-Natta polymerization of isoprene which results in the formation of poly(cis-isoprene), an elastomeric material with a low glass transition temperature. Poly(cis-isoprene) is subsequently treated with one of a variety of reagents under conditions that are proprietary to the manufacturers, but all of which result in partial cyclization of the polymer to produce a material with a higher glass transition temperature and greater structural integrity than its precursor (Figure 13). The cyclized rubber matrix materials are extremely soluble in non-polar,

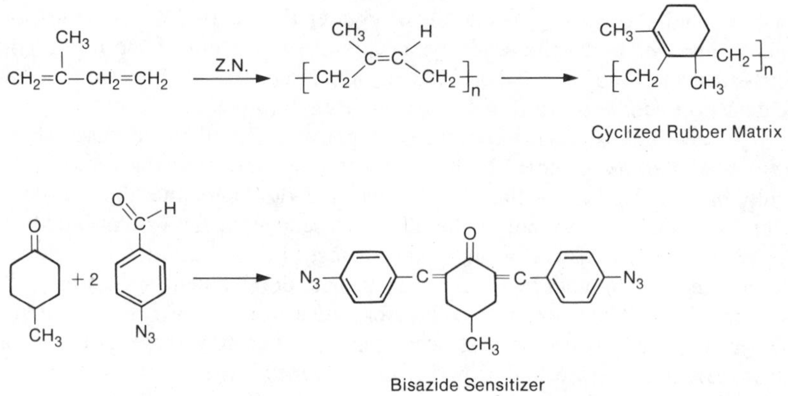

Cyclized Rubber Matrix

Bisazide Sensitizer

Figure 13. Bisarylazide-rubber resists. The matrix resin is cyclized poly (cis-isoprene). The sensitizers are bisarylazides. A typical structure of one commonly employed sensitizer is provided.

organic solvents such as toluene, xylene, or halogenated aliphatic hydrocarbons. The resulting solutions can be spincoated to form isotropic films that adhere strongly to a wide range of substrate materials.

Cyclized rubbers traditionally have been sensitized by a class of molecules that are generically bis-arylazides. A typical structure is provided in Figure 13. The bis-arylazides are synthesized by condensation of para-azidobenzaldehyde with a substituted cyclohexanone. The ultraviolet absorption spectrum of a typical bis-arylazide sensitizer is shown in Figure 14. Note that the sensitizer has a strong absorption centered at approximately 360 nm., and is almost transparent above 420 nm. Across the absorption band, the quantum efficiency for decomposition of the sensitizer is approximately constant. The number of molecules converted per unit time at constant flux varies with absorbance; therefore, the bis-arylazide sensitizer and resist formulated therefrom will be lithographically sensitive to the 365 nm. mercury emission line but extremely insensitive to the 405 and 436 nm. emission lines.

It is possible, through extending the conjugation of the bis-arylazide sensitizers, to shift the absorption maximum to longer wave lengths as depicted in Figure 15. Here the absorption maximum, as well as the lithographic sensitivity, reach a maximum at 400 nm. and are significant at wavelengths up to 450 nm. These "red-shifted" materials are useful for imaging by step-and-repeat exposure tools which typically operate with the

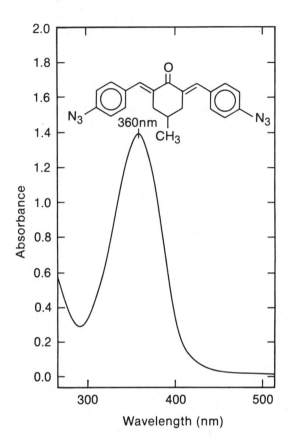

Figure 14. Ultraviolet spectrum of a commonly used bisarylazide sensitizer. The spectrum was measured in solution.

isolated 436 nm. mercury emission line, whereas materials of the sort whose spectra are shown in Figure 14 would have little or no lithographic sensitivity at 436 nm.

Like most negative resists, the arylazide rubber formulations function via the photochemical generation of a crosslinked, 3-dimensional network that is insoluble. The photochemical transformations associated with the generation of this network are depicted in Figure 16. The primary photo event is the evolution of nitrogen from the excited state of the arylazide to

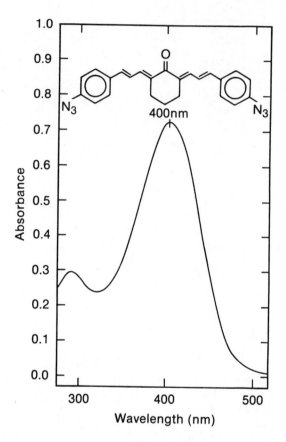

Figure 15. The effect of extended conjugation on the ultraviolet absorbance spectrum of bisarylazides. This spectrum should be compared with Figure 14, the spectrum of a structural analog.

form an extremely reactive intermediate called a nitrene. The reactive nitrene intermediate can undergo a variety of reactions resulting in the formation of more stable structures and in several instances, the generation of a polymer-polymer linkage, or crosslink. The nitrene reactions include nitrene-nitrene coupling to form azo dyes, insertion of the nitrene into carbon-hydrogen bonds to form amines, abstraction of hydrogen from the rubber backbone to form an amine radical and a carbon radical which can subsequently undergo coupling reactions, and finally, insertion of the nitrene into the double bond of the cyclized rubber polymer to form three-membered, heterocyclic aziridine linkages (Figure 16).

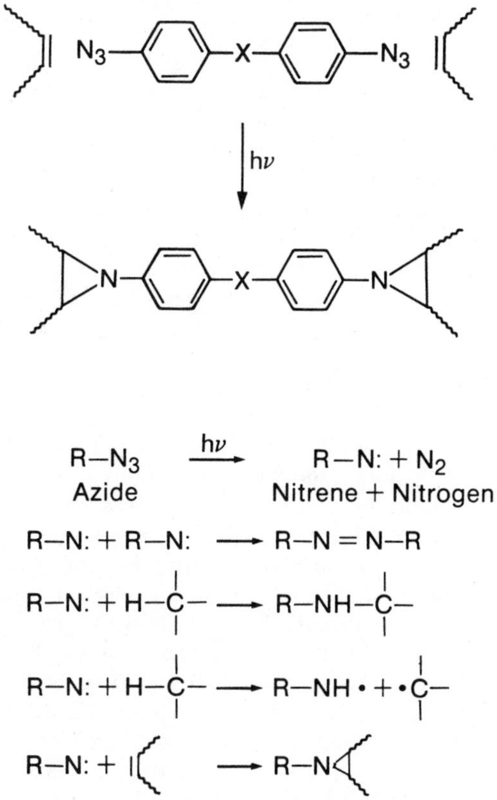

Figure 16. Crosslinking reactions in bisarylazide-rubber resists. The primary photoevent is production of a nitrene which then undergoes a variety of reactions that result in covalent, polymer-polymer linkages. A schematic representation of crosslinking via nitrene insertion to form aziridine linkages is shown together with several other reaction modes available to the nitrene.

3.5.b Two-Component Positive Resists. The bis-arylazide-rubber resist system that was the "workhorse" of the microelectronics industry is slowly being replaced by positive-tone resist systems. In the main, this is occurring because of the higher resolution and thermal stability that positive resists offer and the greater resistance they provide to dry etching environments.

The positive resist materials evolved from discoveries made by the Kalle Corporation in Germany who developed the first positive-acting photoresist based on the use of a novolac matrix resin and a diazoquinone photoactive compound or sensitizer. The original materials were designed to produce photoplates used in the printing industry. These same materials have been adopted by semi-conductor fabrication engineers and continue to function effectively in that more demanding application.

Figure 17 provides an overview of the function of the diazoquinone/novolac materials. The matrix resin is a copolymer of a phenol and formaldehyde. The generic term for this class of polymers is novolac (18) meaning "new lacquer" and describes the purpose for which they were first developed. The chemical industry produces millions of tons of novolac each year where its end use is that of a thermoset resin and adhesive. Novolac is commonly used, for example, as the principle adhesive in the manufacture of plywood.

Novolac resins are soluble in common organic solvents and can be coated from solution to form isotropic, glassy films of high quality. These polymers are also soluble in aqueous base solutions by virtue of the acidic character of their phenolic functionality. The novolac resins that are used in resist formulations are typically of relatively low molecular weight when compared to polymers such as PMMA having number average molecular weight ranging from approximately 300-to-1000 with typically broad dispersities. Common novolacs melt at a temperature of 90 to 120°C and extended heating at temperatures above 130 or 140°C results in the onset of reactions that ultimately result in the crosslinking of the materials.

The photoactive compounds, or sensitizers, that are used in the formulation of positive photoresists, are substituted diazonaphthoquinones shown in Figure 17. The substituent, shown as R in Figure 17, is generally an aryl sulfonate. The nature of the substituent influences the solubility characteristics of the sensitizer molecule and also influences the absorption characteristics of the chromophor (19). The diazonaphthoquinone sulfonates are soluble in common organic solvents but are insoluble in aqueous base. Upon exposure to light, these substances undergo a series of reactions that culminate in the formation of an indene carboxylic acid as depicted in Figure 17. The photoproduct, unlike its precursor, is extremely soluble in aqueous base by virtue of the carboxylic acid functionality.

Positive photoresist formulations consist of a novolac resin and an appropriate diazonaphthoquinone dissolved in organic solvent. Common solvents include: ethyl cellosolve acetate, diglyme, etc. These formulations are spin-coated and then baked to remove the coating solvent. They provide films in which the sensitizer is randomly distributed through the novolac matrix.

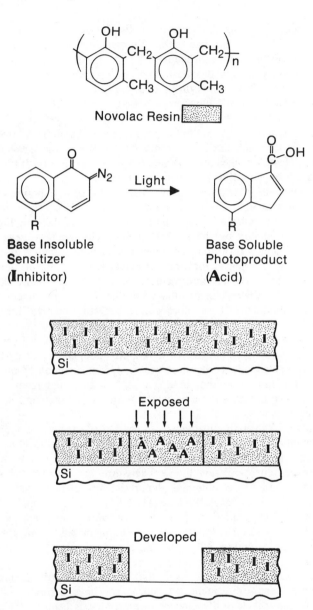

Figure 17. A schematic representation of positive resist action in diazonaphthoquinone-novolac resists. Photolysis of the sensitizer (inhibitor) produces acid which allows the exposed areas of the resist to be selectively dissolved (developed) in aqueous base.

The sensitizer molecule acts as a dissolution inhibitor for aqueous base development of the novolac resin. Films of novolac containing 15- to 20-weight percent of diazonaphthoquinone sensitizer dissolve orders of magnitude more slowly in aqueous base solution than films of novolac alone. Exposure of the sensitized films to light results in conversion of the inhibitor into the base-soluble, acidic photoproduct that increases the dissolution rate of the novolac matrix in the regions where exposure has occurred. Complete photo-decomposition of the inhibitor results in a dissolution rate that is equal to or greater than the intrinsic dissolution rate of the novolac matrix resin alone. This photochemically generated difference in dissolution rate in aqueous base is exploited in the generation of relief images.

The sequence of transformations that occurs upon exposure of diazonaphthoquinones to light has been studied in detail (20). The mechanistic sequences that have emerged from these studies are outlined in Figure 18. Absorption of a photon by the naphthaquinone chromophore generates an excited state which, with a quantum efficiency of approximately 0.2, undergoes evolution of nitrogen to form an extremely reactive intermediate called a carbene. The carbene intermediate undergoes a well-documented sequence of bond rearrangements (known as the Wolff rearrangement) that culminates in production of a more stable but still reactive intermediate called a ketene. The ketene is stable enough to be detected and studied at low temperature, even in a novolac matrix, but at room temperature it rapidly reacts with nucleophiles. In a properly designed resist system, the most reactive nucleophile available is water. Under common humidity conditions, water is always present in the novolac matrix and reacts with the ketene to produce a carboxylic acid which is the final functional product of the sequence of events.

Figure 19 shows the ultraviolet absorption spectrum of a typical diazonaphthoquinone and a common novolac resin. The naphthoquinone sensitizer has a strong absorbance at the 365 nm., 405 nm., and to a lesser extent the 436 nm. mercury emission lines. There are two diazonaphthoquinone isomers that are used in commercial photoresist formulations that are available at this time. The 5-arylsulfonates are by far the most commonly used. A spectrum of a representative of this class of materials is depicted in Figure 20. The 5-arylsulfonate materials are characterized by an absorbance maximum at approximately 400 nm. and a second, slightly stronger maximum at approximately 340 nm.

The 4-sulfonate isomers have found use in certain formulations. These materials are easily identified, since they have a single absorbance maximum in the wavelength region of interest that is centered at approximately 380 nm., and they have little or no absorbance at 436 nm (Figure 21). The absorbance spectra of two representative positive resist formulations are provided in Figures 22 and 23. These spectra were obtained by casting the resists on quartz substrates and then recording the absorbance

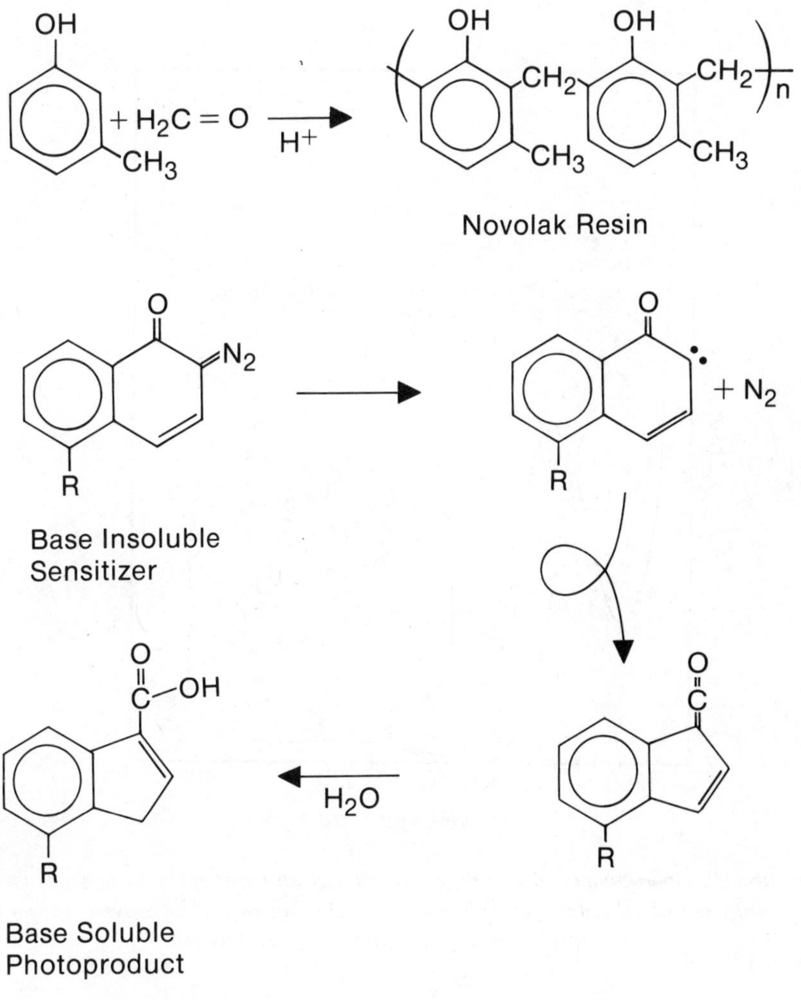

Novolak Resin

Base Insoluble
Sensitizer

Base Soluble
Photoproduct

Figure 18. Diazonaphthoquinone-novolac resist. The novolac (Novolak) matrix resin is prepared by acid catalyzed copolymerization of cresol and formaldehyde. The base insoluble sensitizer, a diazohaphthoquinone, undergoes photolysis to produce a carbene which then undergoes Wolff rearrangement to form a ketene. The ketene adds water which is present in, the film, to form a base soluble, indenecarboxylic acid photoproduct.

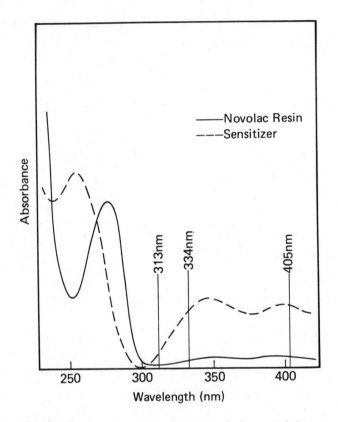

Figure 19. Absorbance spectrum of a typical diazonaphthoquinone sensi-
tizer (in solution) and a cresylic acid novolac (film). The wavelengths of
principle mercury emission lines are labeled.

spectra both before and after exposure to a dose of broad-band uv radiation,
sufficient to totally decompose the sensitizer. AZ1350J (Figure 22) is formulated from a sensitizer that is almost certainly a 5-sulfonate ester.
AZ2300 resist (Figure 24), shows a single broad absorption centered at
approximately 385 nm and must, therefore, be formulated from a 4-
sulfonate analog.

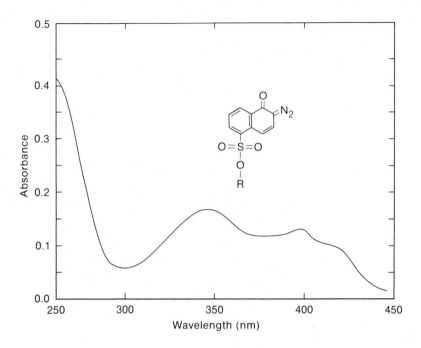

Figure 20. Absorbance spectrum of a 1-oxo-2-diazo-naphthoquinone-5-arylsulfonate. These materials are the most commonly employed sensitizers in positive photoresist formulations.

3.5.c Image Reversal. Before leaving our discussion of diazoquinone novolac resists, it is interesting to consider certain techniques whereby the chemistry of these systems has been permuted in a fashion which allows them to be imaged as high resolution, negative-tone materials. Of particular interest among these is the so-called "monazoline process" (*21,22*). This process involves introducing a small amount of a basic additive such as monazoline (1-hydroxyethyl-2-alkylimidazoline), imidazole or triethanolamine into a diazoquinone-phenolic resin photoresist such as Shipley's AZ1350J, Kodak 820, etc. In the reversal process, the doped resist is exposed through a mask, baked after the exposure, flood exposed, and finally immersed in an aqueous base to generate a high quality, negative tone image of the mask. If the doped resist is simply exposed to uv light and developed in aqueous base in the usual fashion, a positive-tone image is produced. These processes are outlined in Figure 24. The central feature of the mechanism

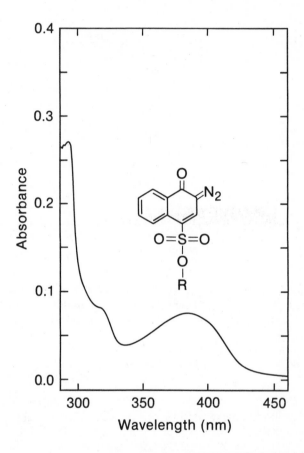

Figure 21. Absorbance spectrum of a 1-oxo-2-diazonaphthoquinone-4-arylsulfonate. These materials differ spectrally from the 5-isomers shown in Figure 20 and are used in certain commercial resist formulations.

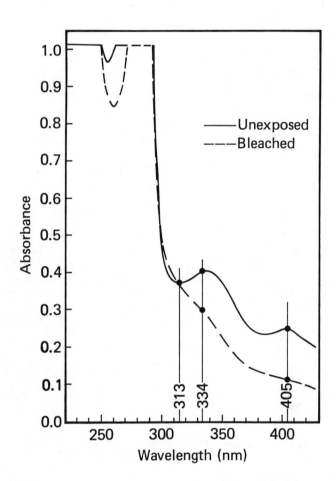

Figure 22. Absorbance spectrum of a 1.17μm film of AZ1350J photoresist coated on quartz. Spectra were recorded before and after exposure.

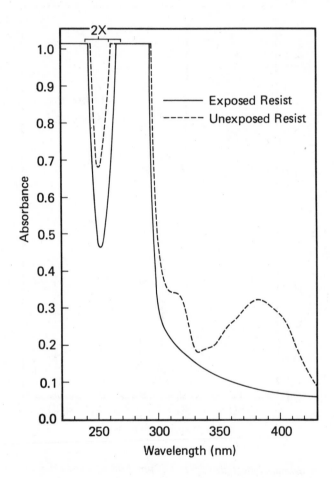

Figure 23. Absorbance spectrum of a 1.05μ film of AZ2400 photoresist coated on quartz. Spectra were recorded before and after exposure.

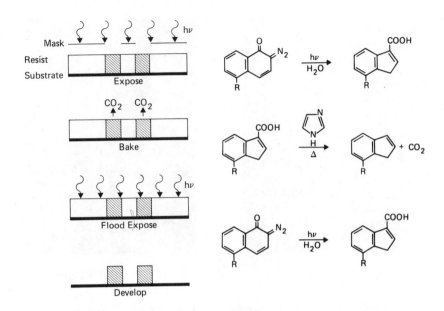

Figure 24. Process sequence for image reversal in a positive photoresist. The chemical transformations of the sensitizer that occur in each process step are provided.

of the monazoline process involves a based catalyzed thermal decarboxylation of the diazoquinone photoproduct (the indene carboxylic acid) during the post-exposure bake step. The decarboxylation product, a substituted indene, is insoluble in aqueous base and is, therefore, a dissolution inhibitor. Flood exposure of the resist after the decarboxylation process destroys inhibitor in the regions of the film where diazoquinone remains such that the final structure prior to development consists of inhibitor (i.e., the indene) remaining only in the areas that were subjected to the image-wise exposure. Consequently, development occurs more slowly in these areas, and a negative-tone image of the mask is generated. This sequence of chemical events is depicted in Figure 24 and represents the results of detailed mechanistic studies carried out at the IBM Laboratories in San Jose and Burlington (23). Scanning electron micrographs of both positive- and negative-tone images generated in imidazole doped AZ1350J resist are provided in Figure 25.

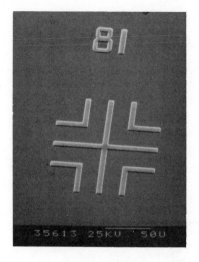
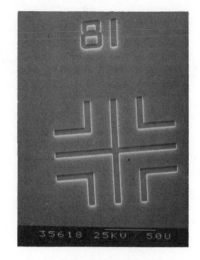

(a) Reversed Image   (b) Positive Image

Figure 25. Electron photomicrographs of AZ1350J resist films doped with imidazole. One film has been subjected to the standard process (b) and the other to the reversal process (a).

3.6 Chemistry of Electron Beam Resist

Poly(methyl methacrylate) (PMMA) is the classical electron beam resist and offers the advantage of extremely high resolution, ease of handling, excellent film forming characteristics, wide processing latitude, and ready availability. It is, unfortunately, a relatively insensitive material requiring 50 to 100 $\mu C/cm^2$ at 20 kV exposure dose in most applications. This lack of sensitivity and the corresponding impact it has upon production throughput, together with the fact that PMMA is a less efficient plasma or reactive ion etch mask than resists constituted from aromatic materials such as the novolacs (24) (Figure 26) has prompted a great deal of research directed toward the development of more sensitive electron beam resist materials that incorporate reactive ion etch resistance.

The resist materials that have emerged from these studies are the first examples of resist materials tailored specifically for use in the microelectronics industry. They are specialty polymers that are produced in very low volumes, and have been developed as a result of research carried out in industrial laboratories. In the following sections, we will examine the chemistry of selected examples of these new systems.

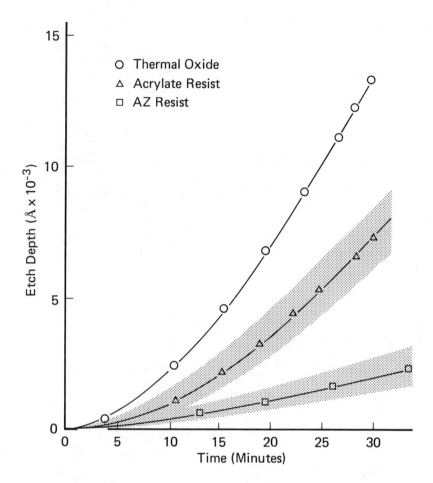

Figure 26. A comparison of the etch rates of thermal oxide, acrylate resist (PMMA) and AZ resist using DE-100 gas at 200W and 0.55 Torr. The shaded areas surrounding the acrylate and AZ curves represent the etch rates of typical aliphatic and aromatic polymers respectively.

3.6.a The Acrylates. PMMA is produced by radical initiated polymerization of commercially available methyl methacrylate monomer. Dupont's Evacite 2041 and 2010 are widely available and have become the standard materials for laboratory e-beam and x-ray experiments. PMMA has a $G(s)$ value of 1.3 and has no measurable crosslinking propensity. The radiochemical conversions that culminate in main chain scission have been studied in detail by several workers using a variety of sophisticated analytical techniques (25-30). The initial radiochemical event appears to be homolysis of the main chain carbon to carbonyl carbon bond or homolysis of the carbonyl carbon to oxygen sigma bond. In the latter case, homolysis is followed by rapid decarbonylation, to form the same, stable tertiary radical on the main chain, as depicted in Figure 27. The main chain radical undergoes rearrangement through beta scission to cleave the chain and generate an acyl-stabilized, tertiary radical as indicated. This process generates fragments of carbon monoxide, carbon dioxide, and methyl and methoxyl radicals.

Attempts to improve the sensitivity of PMMA through synthesis of analogs while preserving its attractive processing characteristics occupied resist chemists for several years during the early 70's and research in the area of acrylate radiation chemistry continues to produce new results. The first electron beam resist used in device manufacturing can be considered to

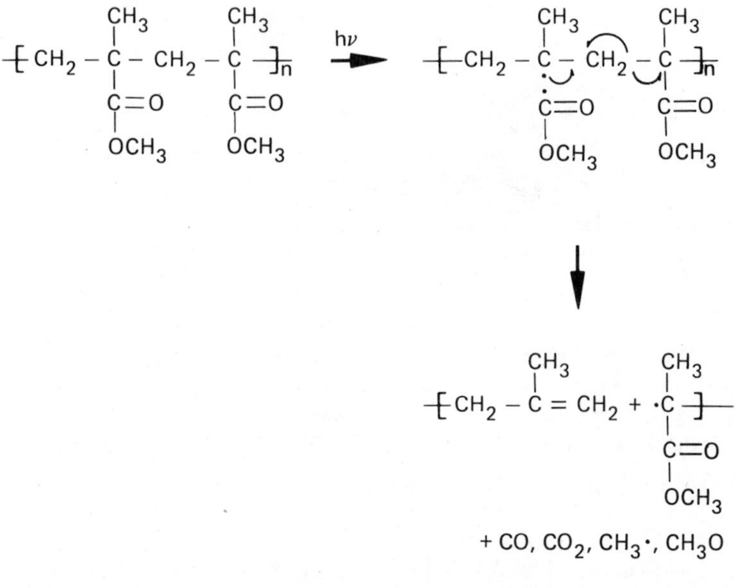

Figure 27. Mechanism of radiation induced chain scission in PMMA. Homolysis of the mainchain-carbonyl carbon bond is indicated as the initial step. Acylcarbon-oxygen, sigma bond homolysis also occurs but rapid decarbonylation ultimately leads to the same indicated products.

have evolved from research in the area of acrylate radiation chemistry. Moreau and coworkers at IBM have described a resist material called "Terpolymer" that was introduced into manufacturing in IBM during the mid 70's (*31*). "Terpolymer" resist (Figure 28) is prepared by copolymerizing methyl methacrylate and methacrylic acid to produce a copolymer which is subsequently heated to generate intramolecular anhydride linkages such that the final product is, in fact, a polymer consisting of three unique monomeric units (a terpolymer), methyl methacrylate, methacrylic acid and methacrylic anhydride. The G(s) of "Terpolymer" is approximately 4.5 scissions per 100 eV, and it has a useful resist sensitivity, at 20 kV, of approximately 10 $\mu C/cm^2$ as practiced in manufacturing.

"Terpolymer" resist clearly demonstrates a significant improvement in sensitivity over PMMA. In addition to the changes in structure that are represented in "Terpolymer" resist, a wide range of PMMA analogs has been reported and all have increased resist sensitivity over PMMA. These include contributions from industrial laboratories in the U.S., Europe and Japan, as well as those of academic researchers (*32*). A compilation of representative examples of these materials and their radiochemical G values and lithographic sensitivities are provided in Figure 29. Of these materials, only FBM, a poly(fluorobutyl methacrylate) and certain fluroalkyl analogs thereof are commercially available (*33-35*).

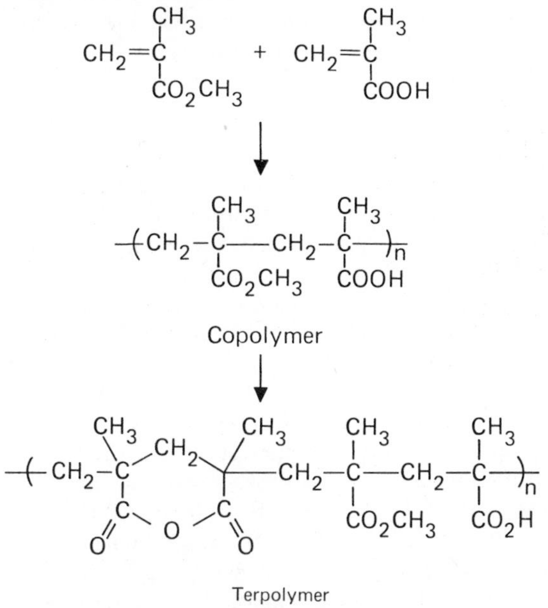

Figure 28. *"Terpolymer" e-beam resist.*

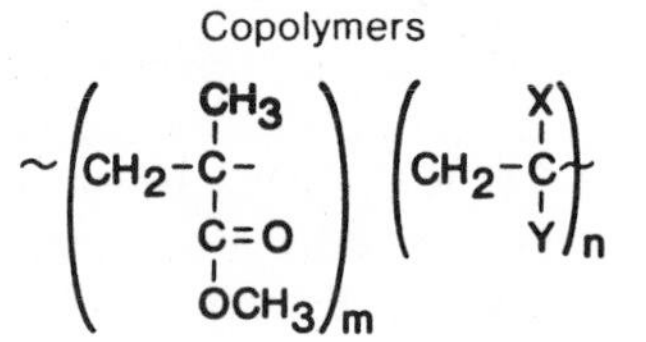

Copolymers

$$\sim \left(\begin{array}{c} CH_3 \\ | \\ CH_2-C- \\ | \\ C=O \\ | \\ OCH_3 \end{array} \right)_m \left(\begin{array}{c} X \\ | \\ CH_2-C \\ | \\ Y \end{array} \right)_n$$

Copolymer	(15 Kv) Sensitivity, $\mu coul/cm^2$	G_S
PMMA	40	1.5
PMMA-MA (X=CH$_3$, Y=COOH)	35	2.0
PMMA-MAN (X=CH$_3$, Y=CN)	12	3.1
PMMA-IB (X=CH$_3$, Y=CH$_3$)	14	3.5
PMMA-α CL-Acrylate (X=CL)	14	3.3
PMMA-α CR-Acrylate (X=CN)	12	3.5
PMMA-MA-MANH	7	4.5

Figure 29. The 15 kV e-beam sensitivity and the γ-ray G (s) of a variety of PMMA analogs. Note that increasing G (s) is inversely related to the lithographic sensitivity.

3.6.b The Poly (olefin Sulfones). Poly(olefin sulfones) are typically alternating one to one copolymers of sulfur dioxide and an alkene and have been known for almost 100 years. Solonia observed in 1898 that allyl ethers react with sulfur dioxide to produce a white amorphous material (*36*). It was not until 1934 that the polymeric nature of this amorphous material and its one to one alternating structure were elucidated by Marvel and coworkers (*37*). Since that time, several studies on the synthesis and physical characteristics of poly(olefin sulfones) have been published by both industrial and academic laboratories.

An early commercial interest in poly(olefin sulfones) was sparked by the low raw materials cost, but this interest waned when it became apparent that thermal instability is a general characteristic of this class of materials. In 1970 Brown and O'Donnell reported that poly(butene-1-sulfone) is degraded by gamma radiation with a G(s) approaching 10, making it one of the most radiation-sensitive polymers known (*38-39*). The potential for use of this radiation sensitivity in the design of electron beam resists was quickly realized by several members of the electronics industry. Bell Laboratories, RCA, and IBM published studies demonstrating the potential of poly(alkene sulfones) as resists for electron beam lithography. However, the transformation of this technology from a laboratory curiosity into a manufacturing reality required the ability to routinely prepare lithographically useful samples of poly(alkene sulfone). To the best of our knowledge, the only lithographically useful poly(olefin sulfone) that has been produced on a commercial scale is poly(butene-1-sulfone), PBS, which was developed at Bell Laboratories (40).

PBS (Figure 30) is an alternating copolymer of sulfur dioxide and 1-butene. It undergoes efficient main chain scission upon exposure to electron beam radiation to produce, as major scission products, sulfur dioxide and the olefin monomer. Exposure results first in scission of the main chain carbon-sulfur bond, followed by depolymerization of the radical (and cationic) fragments to an extent that is temperature dependent and results in evolution of the volatile monomers species. The mechanism of the radiochemical degradation of polyolefin sulfones has been the subject of detailed studies by O'Donnell et. al. (41).

PBS is a far more sensitive resist material than PMMA as one would predict by comparison of their relative G-values. In that sense, discovery of the radiation sensitivity of this class of polymers has succeeded in providing the order of magnitude greater sensitivity that resist chemists had sought. The sulfone polymers are, however, sensitive not only to electron beam radiation, but to plasma and reactive ion etching environments. PBS, for example, can be successfully imaged at doses of less than $1 \ \mu C/cm^2$ at 10 kV. The resulting images are useful for wet etching but have limited utility as dry etch masks. PBS has found wide applicability in electron beam mask making but does not provide a solution to the problem of a high sensitivity, dry etch resistant e-beam resist for use in direct-write wafer manufacturing.

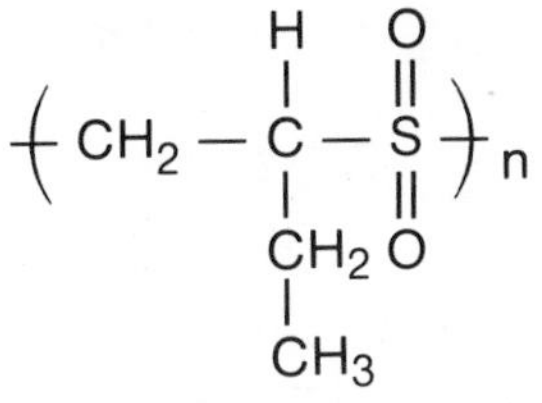

$$\left(CH_2 - \overset{\overset{\displaystyle H}{|}}{\underset{\underset{\displaystyle CH_2}{|}}{C}} - \overset{\overset{\displaystyle O}{\|}}{\underset{\underset{\displaystyle O}{\|}}{S}} \right)_n$$
$$\underset{CH_3}{|}$$

$$R-SO_2-R' \longrightarrow [RSO_2R']^{\overset{\cdot}{+}} + e^-$$

$$[RSO_2R']^{\overset{\cdot}{+}} \longrightarrow RSO_2^+ + \cdot R' \rightarrow R^+ + SO_2$$

Figure 30. Poly (butene-1-sulfone), PBS. The radiochemical decomposition pathway via a radical cation is proposed by O'Donnell.

The problem of etch resistance in the application of sulfone polymers to electron beam resist design seems to have been circumvented in the development of NPR (New Positive Resist), the second generation electron beam resist material that has evolved from work at Bell Laboratories (*42*).

NPR consists of a novolac matrix resin, similar to that used in the common diazoquinone-novolac photoresist materials, and a poly(olefin sulfone) sensitizer (Figure 31). This two-component positive resist appears to provide at least a factor of two greater sensitivity than terpolymer resist combined with significantly higher resistance to reactive ion etching and plasma environments. The design of this resist system succeeds in expressing the favorable characteristics of the two components from which it's constituted, viz., the excellent film forming characteristics and etch resistance of the novolac polymers and the high radiation sensitivity of the poly(olefin sulfone) sensitizer material.

3.6.c The Epoxy Resists. The first negative tone electron beam resist materials with useful sensitivity were based on utilizing the radiation chemistry of the oxirane or epoxy moiety. The most widely used of these materials, COP (Figure 32) is a copolymer of glycidyl methacrylate and ethyl acrylate and was developed at Bell Laboratories (*43,44*). COP has found wide applicability in the manufacturing of photomasks. The active element

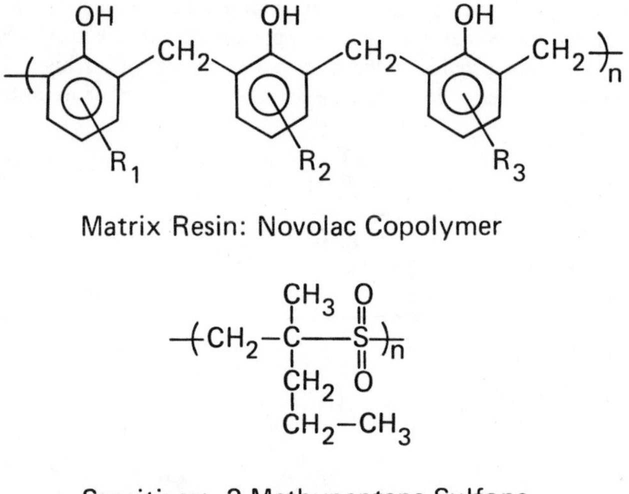

Matrix Resin: Novolac Copolymer

Sensitizer: 2-Methypentene Sulfone

Figure 31. NPR, Bell Laboratories "New Positive Resist" for e-beam applications. The sensitizer is a radiolabile polysulfone and the matrix resin is a novolac.

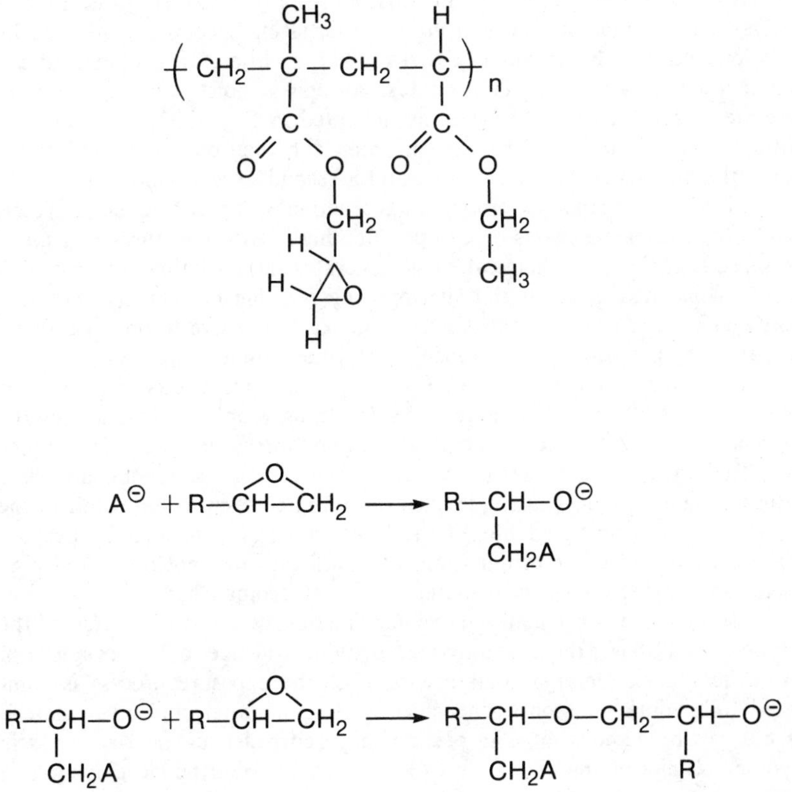

$$A^{\ominus} + R-\underset{\diagdown O \diagup}{CH-CH_2} \longrightarrow R-\underset{\underset{CH_2A}{|}}{CH}-O^{\ominus}$$

$$R-\underset{\underset{CH_2A}{|}}{CH}-O^{\ominus} + R-\underset{\diagdown O \diagup}{CH-CH_2} \longrightarrow R-\underset{\underset{CH_2A}{|}}{CH}-O-CH_2-\underset{\underset{R}{|}}{CH}-O^{\ominus}$$

and other mechanisms

Figure 32. COP, Bell Laboratories negative e-beam resist. The resist is a copolymer of glycidyl methacrylate and ethyl acrylate.

in the structure of COP is the glycidyl side chain which carries the three-membered heterocyclic substituent called an oxirane or epoxy group. This is the same active element that is found in the classical epoxy adhesives with which we are all familiar.

COP functions on the basis of radiation-initiated crosslinking reactions that result in the formation of interchain linkages which generate a crosslinked, three-dimensional network that is insoluble. The chemistry associated with the generation of this network is interesting in that it involves a chain reaction mechanism. The initiating species is generated by radiation and may be an anion, anion radical, a cation, or a cation radical. The exposure-generated initiator attacks an epoxy moiety and cleaves it to generate a reactive oxygen species, as indicated in Figure 32, for the example of anionic initiation. The oxygen anion can then open a second epoxy group, thereby generating a crosslink and a second oxygen anion which may be involved in yet another crosslinking reaction. The consequence of this chain propagation mechanism is to produce high sensitivity through *gain* in the sense that the radiochemical event generates the initiating species which may be consumed in the initial ring opening reaction but which simultaneously generates a product that is itself sufficiently reactive to produce a propagating, chain reaction of ring-opening polymerization sequences.

The measured $G(x)$ value of representative epoxy polymers is approximately 10, but this value depends strongly on the structure of the polymer, its glass transition temperature and other characteristics. Since the crosslinking reaction that characterizes the COP resist functionality is a chain reaction, in theory, a single, electron-initiated event could result in the insolublization of an entire film of the resist material. Fortunately, because of the existence of chain terminating reactions, this does not occur and high resolution imaging of the resist material can be accomplished.

In negative resist materials which function by a chain reaction of the sort described above, there is a growing body of evidence for the existence of a so-called "dark reaction" that occurs after the exposure process is completed. This dark reaction is quenched by oxygen, water, and other materials but can be demonstrated to occur if exposed resist material is stored in vacuum. A plot of the change in exposed feature size after developing as a function of vacuum curing time for several resist materials that incorporate gain is provided in Figure 33 (*45,46*). The dark reaction serves to increase the apparent sensitivity of these materials to e-beam radiation but offers the disadvantage of a time dependence on the size of developed features.

3.6.d The Styrene Resists. Polystyrene (Figure 34) is a common industrial plastic. The polymer itself is a weakly sensitive, negative electron resist. However, "minor" perturbations in the structure of polystyrene have a profound effect on its radiation chemistry. Poly(α-methylstyrene), for example, is a weakly sensitive, *positive* electron beam resist with a $G(s)$ of less than 1 but with unmeasurable $G(x)$. Substitution of polystyrene on the ring, in particular with halogen or halomethyl groups (Figure 34), has produced a new generation of negative electron beam resists that exhibit very high sensitivity and appear to have no dark reaction (Figure 35) (*47-50*). The absence of a dark reaction is consistent with the proposed mechanism of action which involves radiation-induced homolysis of the carbon-halogen bond to generate a radical which then may undergo rearrangement, abstrac-

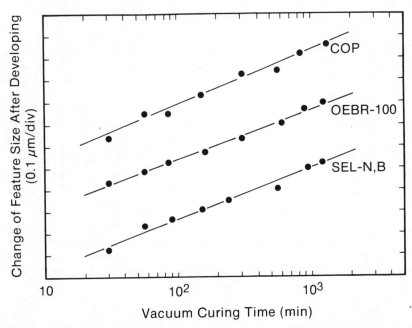

Figure 33. The change in developed line-width with vacuum storage time after exposure for three commercially available, negative e-beam resists.

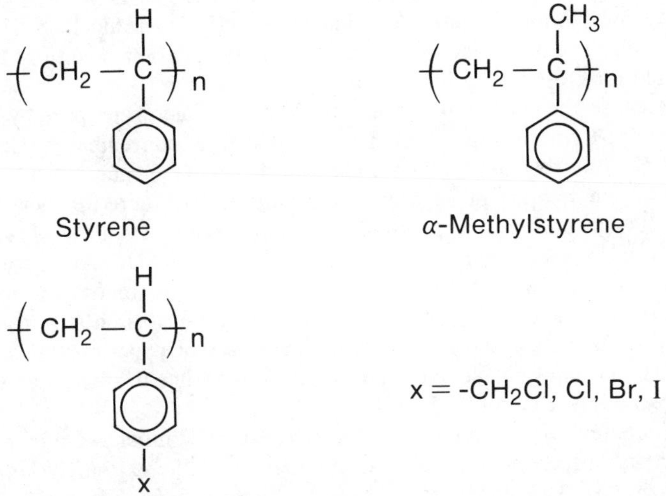

Styrene

α-Methylstyrene

$x = $ -CH$_2$Cl, Cl, Br, I

Figure 34. The Polystyrenes. Polystyrene is a low sensitivity negative resist, poly(α-methylstyrene) is a low sensitivity positive resist and the para-substituted analogs listed are all sensitive negative resists.

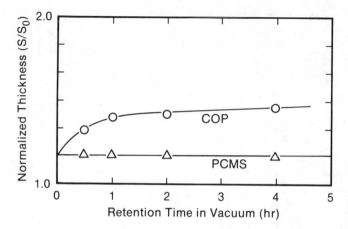

Figure 35. The normalized change in thickness remaining after development vs storage time after exposure for COP resist and poly (chloromethylstyrene). The styrene analog shows little or no "dark-reaction".

tion, or recombination reactions that ultimately lead to network formation. There is no obvious gain mechanism in these reactions; therefore the high sensitivity must imply that the G value of the primary radiochemical event is much higher than the corresponding radio-initiation reaction that is responsible for crosslinking in the epoxy materials.

Because the styrene polymers are relatively easy to prepare by classical radical polymerization, they have provided a convenient vehicle for studying the effect of molecular weight and dispersity on the sensitivity and contrast of one component negative resist materials. Figure 36 shows sensitivity plots generated for poly(chloromethylstyrene) samples of varying molecular weight but essentially constant dispersity (47). An increase of molecular weight by a factor of 10 results in approximately a ten-fold increase in resist sensitivity, while the contrast remains essentially constant. Figure 37 is a plot of the results of a similar set of experiments that was carried out on poly(p-chlorostyrene) samples differing in molecular weight by approximately a factor of 2 (50).

The effect of reducing dispersity at constant number average molecular weight is to increase the contrast (γ) (51-53). The extremely high contrast of the poly(p-chlorostyrene) samples described in Figure 37, is the result of having fractionated these materials such that their dispersities are much less than 2. The general trend is that increased molecular weight at

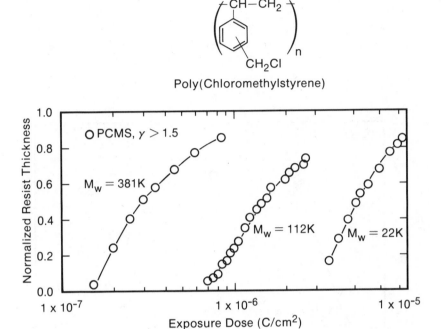

Poly(Chloromethylstyrene)

Figure 36. Effect of change in molecular weight at constant dispersity on the e-beam sensitivity of poly(chloromethylstyrene), PCMS. (Reproduced with permission from Ref. 47.)

constant dispersity results in increased sensitivity. Decreased dispersity at constant number average molecular weight results in increased contrast to some limit that is structure dependent.

One aspect of negative resist performance that cannot be inferred from an examination of the thickness remaining vs. log exposure plots such as Figures 36 and 37 is the influence of swelling on image acuity. It is certainly true that resist materials with high contrast (γ) have a potential for producing steep sidewall images and therefore high resolution. Unfortunately, these images may be distorted by swelling during development. The swelling phenomonon is a difficult one to avoid, as it results from the fact that any solvent which would dissolve the polymer in the unexposed areas of the resist film must necessarily associate itself with that polymer

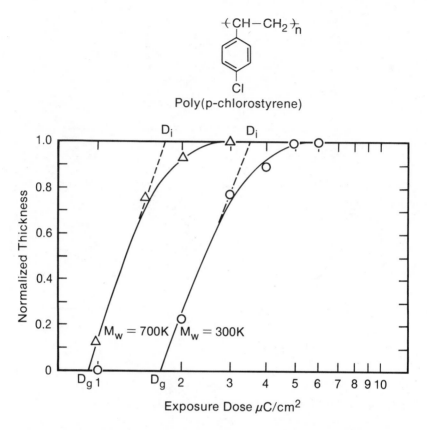

Figure 37. The effect of change in molecular weight at constant dispersity on the e-beam sensitivity of poly(p-chlorostyrene). The very high contrast was achieved by fractionating to obtain nearly monodispersed samples.

structure to produce a lower energy state than the energy of the polymer/polymer interaction itself (see Chapter 4) (54).

During the development process, the cross-linked area absorbs a large volume of solvent and must necessarily, therefore, increase its dimensions. This increase in volume, termed swelling, causes image distortion. In particular, closely spaced lines may swell to such an extent during the development process that adjacent networks may join. Upon removal of solvent, the network shrinks but "bridging" occurs between the lines, as a result of material being transferred from one line to its adjacent neighbor.

The other common manifestation of the swelling phenomenon is the formation of snake-like distortions of long, narrow images. "Snaking" results from the fact that during development it is possible for these structures to expand in the vertical direction, but because of their adherence to the substrate, they cannot expand freely in the plane of the substrate. The dilemma, then, is that these long, narrow lines must increase their length to relieve stress but yet cannot change their linear dimension without losing adhesion to the substrate. Resolution of the problem is achieved by generation of a sinusoidal, snake-like appearance that often persists even after drying, as evidenced in Figure 38.

It should be noted that useful, high resolution (0.75 μm) patterns can be produced in certain negative resists if the development step is followed by a sequene of rinses in solvents that have less and less affinity for the polymer structure. The rinse steps effectively reverse the swelling that occurs during development.

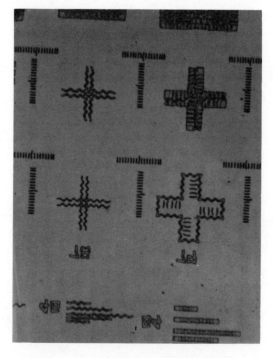

Figure 38. Optical photomicrograph of swollen images in an experimental, negative optical resist demonstrating the "snaking" phenomenon. Stress relief in long, narrow resist lines is achieved by the obvious, oscillatory distortion.

3.6.e Nonswelling Negative Resist Systems. It is evident from the previous section that high resolution imaging can be accomplished in negative e-beam resist materials if these materials have a high contrast (γ) and if distortion due to swelling can be minimized. High γ can be achieved by carefully controlling the dispersity of the polymer through judicious choice of the synthetic conditions or by fractionation. Solution of the problem of potential resolution loss due to swelling required a new insight. Hofer and coworkers, in 1980, described a new resist system that is negative in tone, has high contrast, and is devoid of swelling (*55*). This new resist material, which they term PSTTF (Figure 39) incorporates a new conceptual design for negative resist action. The differential dissolution rate is achieved not through generation of a three-dimensional, crosslinked network, but rather through an alteration of the chemical nature of the polymer side chain that results in a large change in the polarity and, therefore, solubility characteristics of the material in the exposed areas.

PSTTF is based on polystyrene that has been functionalized through appendage of a tetrathiofulvalene side-chain. The resist is spin-coated from a solution which contains a sensitizer substance, typically a perhaloalkane

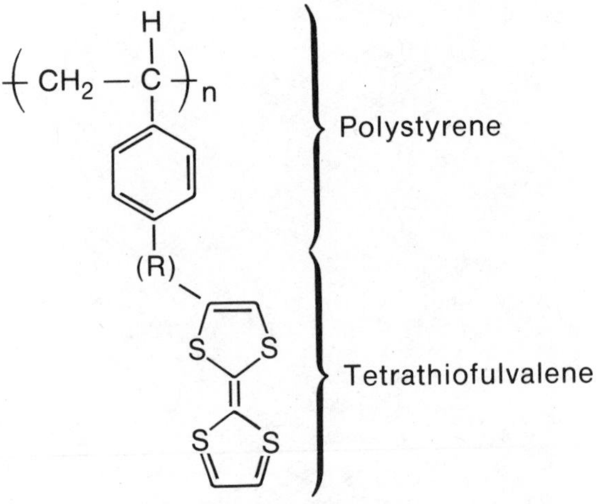

$$PSTTF + CBr_4 \xrightarrow{h\nu} PS[TTF]^+ Br^{\ominus}$$

Figure 39. Polystyrene-tetrathiofulvalene resist. The sensitizer is a perhaloaliphatic such as CBr_4. Exposure results in salt formation in the exposed areas of the resist film.

such as carbon tetrabromide. Exposure of the resist films results in the generation of tetrathiofulvalene bromide salt in the exposed areas of the film. The salt is extremely polar in nature and is, therefore, insoluble in common organic solvents. Consequently, development of the exposed wafer with a nonpolar solvent selectively dissolves the nonpolar poly(styrene tetrathiofulvalene) polymer but does not interact with the polar, exposed areas in any substantial way.

Scanning electron micrographs of high resolution negative resist images generated in PSTTF resist (Figure 40) show nearly vertical resist sidewalls and no discernible distortion due to swelling. In a later section, we will see two recent examples of application of the design concept embodied in PSTTF resist, that is, differential solubility generation through alteration of the polarity of polymer sidechain groups rather than differential solubility generated through backbone scissioning reactions or crosslinking phenomena.

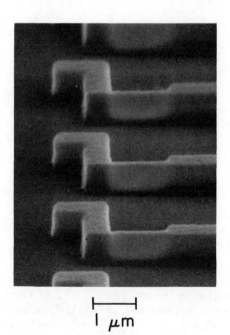

⊢——————⊣
I μm

Figure 40. Scanning electron micrograph of x-ray exposed PSTTF resist.
The vertical wall profiles are indicative of high contrast and there is no
evidence of distortion due to swelling.

3.7 X-ray Resists

Resist materials that are sensitive to electron beam radiation are also sensitive to x-ray radiation and function in the same fashion, that is, materials that are positive in tone for electron beam radiation are typically positive in tone for x-ray radiation. In fact, to first order, there is a strong correlation between the sensitivity of resist systems (positive or negative) to electron beam radiation and their corresponding sensitivity to x-ray radiation (56). Figure 41 is a plot of the 20 kV electron beam sensitivity in C/cm^2 plotted against the Mo soft x-ray sensitivity in mJ/cm^2 of incident x-ray flux. This plot graphically demonstrates the high correlation between e-beam and X-ray sensitivity as evidenced by the unit slope of the plot. The conclusion that must be drawn from analysis of such data is that the basic radiation chemistry that these materials undergo is the same under exposure to both

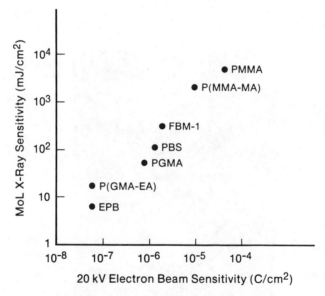

Figure 41. A plot of sensitivity to Mo L_α (5.4Å) x-ray radiation and 20 kV electron beam radiation for several resists. EPB is epoxidized polybutadiene, P(GMA-EA) is a copolymer of glycidyl methacrylate and ethyl acrylate (COP), PGMA is poly(glycidyl methacrylate), PBS is poly(butene-1-sulfone), FBM-1 is poly(2,2,3,3-tetrafluoropropyl methacrylate), P(MMA-MA) is a copolymer of methyl methacrylate and methacrylic acid, PMMA is poly(methyl methacrylate). (Reproduced with permission from Ref. 56.)

forms of radiation. A similar analysis comparing electron beam sensitivity to ion beam sensitivity (57) is provided as Figure 42. The same sort of correspondence holds, and the same conclusion must be drawn regarding the mechanisms that are operative.

A closer analysis of Figures 41 and 42 provides some additional insights. The offset in the axis of Figure 42 which correspond to approxi-

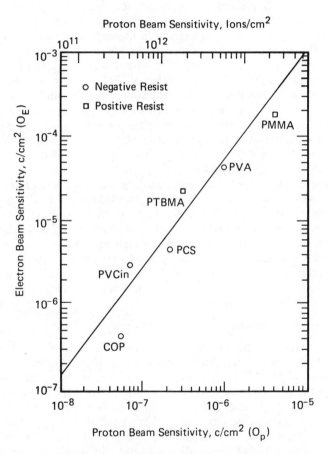

Figure 42. A plot of proton beam vs electron beam sensitivity for several resist systems. COP is a copolymer of glycidyl methacrylate and ethyl acrylate, PVC is poly (vinylcinnamate), PCS is poly (chlorostyrene), PTBMA is poly (t-butyl methacrylate), PVA is poly (vinyl acetate), PMMA is poly (methyl methacrylate). (Reproduced with permission from Ref. 57.)

mately a decade of difference in sensitivity between proton and electron beam exposure, can be readily explained in terms of the increased cross section for absorption that organic materials offer to protons than to electrons. A plot of the resist sensitivities based on energy absorbed per unit volume rather than incident dose serves, within experimental error, to normalize the axis systems.

The increase in apparent sensitivity that results from more efficient absorption of incident flux has been discussed in detail in an earlier section on optical resists. Application of this principle can also be applied to x-ray resists and succeeds in placing certain materials above the line shown in Figure 41. The increase in absorption is accomplished by matching the characteristic x-ray emission wavelength of the exposure source to the absorption edge of an element used in the construction of a resist material. Examples of this wavelength matching include introduction of chlorine into materials designed for exposure to palladium emissions and fluorine into resists designed for exposure with aluminum K_α radiation. Maximal effectiveness is achieved when the halogen atom is introduced into the resist material in a way that is rational with respect to the generation of those intermediates that lead to scission in the case of positive resists, or crosslinking, in the case of negative resists.

An example of the wavelength matching technique is apparent in the work of Taylor et. al. (58,59). Taylor and coworkers at Bell Laboratories have demonstrated very high sensitivity in 2,3-dichloropropyl acrylate-based resist systems for exposure to the palladium emission line. The sensitivity of these materials is in part the result of the high absorption cross section of chlorine for the palladium radiation. With the exception of apparent sensitivity perterbations that can be explained on the basis of unique absorption characteristics, there seem not to be new principles involved in the design of resist materials for ion beam or x-ray exposure.

3.8 Dry Processed Resist

With the advent of plasma and reactive ion etching procedures for the transfer of resist images into semi-conducting and insulating materials, there logically evolved an effort on the part of resist chemists to utilize these new techniques in the generation of the primary resist relief images. The reported advantage of using dry processing to obviate the chemical development steps in lithographic relief image formation are 1) a reduction in the amount of organic waste material that must be processed by a semi-conductor manufacturing plant, 2) a reduction in defect density that is expected to occur as a result of the fact that the device structures would be handled in a vacuum system rather than in a solvent bath and 3) eliminating swelling during development.

The first example of resist relief image formation in the absence of a wet developing step was reported by Bowden and co-workers in 1974 (*60*). These workers reported that exposure of certain poly(olefin sulfones) to electron beam radiation resulted in spontaneous relief image formation. If the films were cast thin enough and the substrates were heated, it was possible to produce clean images in the resist films by exposure alone, thereby avoiding a wet development step.

In 1979, Smith and co-workers described the development of a system they called PDP (which presumably stands for Plasma Developable Photoresist) that is based on the use of a material, the structure of which has not yet been divulged (*61*). In this process the resist is coated in the usual fashion and exposed optically. The exposed film is then subjected to a baking cycle that produces a relief image of negative-tone; that is, depressions are generated in unexposed areas (Figure 45). This relief structure is

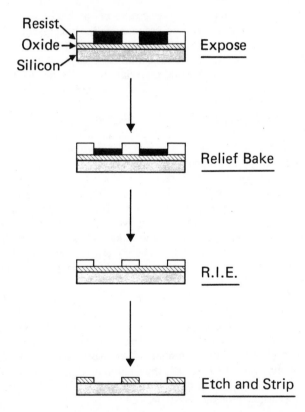

Figure 43. A schematic representation of the PDP, plasma developed resist process.

then transferred into the underlying substrate by reactive ion etching to produce a negative-tone relief image of the mask.

More recently an x-ray resist system that functions in a fashioned analogous to PDP has been described in detail (62,63). This system is based on the use of poly(2,3-dichloropropyl acrylate) as a host polymer and the addition of an organometallic monomer, generally a silicon-containing material, that is characterized by the inclusion of a reactive acrylate functionality as indicated in Figure 46. Exposure of films containing the polymer and this organometallic, functional monomer results in radical-initiated crosslinking of the polymer in the usual fashion, except that the reactive monomer unit is incorporated into this network through inclusion of one or more of its reactive sites. The resulting films are then heated such that the organometallic monomer is removed by volatilization except in those areas where, through radiochemical reaction, it has been covalently incorporated into the film.

The developing step involves exposure of the baked resist films to an oxygen plasma. Interaction of the organometallic moiety with this plasma generates a refractory metal oxide that acts as an etch barrier in those areas in which it is still present within the film (the exposed area) whereas in areas where it's absent, the organic material is rapidly removed by the oxygen plasma. The consequence of this process is generation of negative-tone relief images of the mask.

The general concept embodied in this design has been effectively utilized by Japanese workers in the development of a dry processed optical resist that is based on the use of bis-arylazides and an aliphatic matrix resin such as poly(methyl isopropenyl ketone) (64-65). Exposure of the resist results in nitrene formation as described previously (Figure 16) and consequent covalent bonding of the aromatic materials into the aliphatic matrix network. After exposure, the resist films are heated and the undecomposed arylazide is driven from the film. Negative-tone relief image formation is achieved by plasma development of the baked films in an appropriate gas mixture chosen to optimize the difference in etch rate between aliphatic and aromatic organic compounds. The image quality that has been achieved in this process is excellent.

The latest addition to this list of dry developing resist materials is a contribution from IBM's San Jose Research Laboratory (66-67) that evolved from efforts to design positive-tone resist materials that incorporate *chemical amplification*. These efforts were stimulated by the fact that the quantum yield of typical diazoquinones of the sort used in the formulation of positive photoresists is 0.2 to 0.3; thus, three or four photons are required to transform a single molecule of sensitizer. This places a fundamental limit on the photo-sensitivity of such systems.

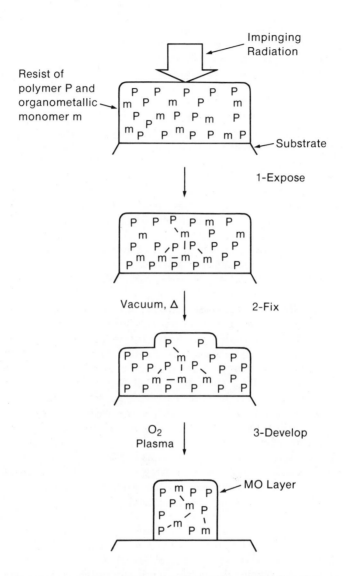

Figure 44. A schematic representation of the plasma developed x-ray resist process. Exposure serves to covalenty bind the monomer (m) into the polymer matrix (p). Heating (fixing) drives out (volatilizes) the monomer except where it is "locked in place" by exposure. Plasma treatment converts the silicon to SiO_2 which retards the etch rate in the exposed areas through formation of a metallic oxide (MO) layer.

In order to circumvent this sensitivity limitation, the San Jose researchers sought to design resist materials that incorporate *chemical amplification* of the sort that characterizes the silver halide photographic emulsion system. In these systems a single photo event initiates a cascade of subsequent chemical reactions that ultimately result in the intended function.

A new positive-tone, dry developing material resulted from successfully designing two stages of amplification into a positive resist material. The polymer upon which the system is based is a polyaldehyde. Aldehydes undergo anionic polymerization in a reversible equilibrium reaction. Many of these systems have ceiling temperatures well below room temperature and the polymerization reactions must be run at cryogenic temperatures. If the polymer is isolated or the reaction is allowed to warm up, the product rapidly depolymerizes to monomer. If, however, these polymers are end-capped by acylation or alkylation prior to isolation or warming, they are often quite stable. Most of the polyaldehydes are intractable, insoluble substances because of their high crystallinity. However, certain aromatic dialdehydes such as phthalaldehyde and o-formylphenylacetaldehyde undergo cyclopolymerization below their ceiling temperatures of approximately -40°C. After end-capping, these materials are stable to greater than 150° and are soluble in common organic solvents (*68*). A formulation consisting of end-capped polyphthalaldehyde, sensitized by addition of cationic photoinitiators, such as triarylsulfonium or diphenyliodonium metal halides (*66-67*), has allowed imaging of 1-μm thick films of the resist formulation at extremely low doses to produce clean relief patterns with resolution below 1-μm linewidth in which the relief pattern is generated spontaneously upon exposure and does not require development by solvent, plasma or heating.

The mechanism by which this system is thought to function involves photogeneration of a strong acid which catalyzes cleavage of the polyacetal main chain. Since the acid is not consumed in the acidolysis reaction, a single molecule of acid is capable of cleaving many chains, thus producing one stage of gain or amplification in the system. The second stage of amplification is achieved because a single cleavage of the polyacetal chain re-establishes the polymerization equilibrium at a temperature above the ceiling temperature, and the polymer simply "unzips" into monomeric fragments. Apparently, under the conditions of the exposure, the monomer is volatilized and spontaneous relief image formation results.

Films of polyphthalaldehyde, sensitized by cationic photoinitiators, have been imaged at 2-5-mJ/cm^2 in the deep ultraviolet (DUV) (see Section 3.10), at 1 $\mu C/cm^2$ (20 kV) electron beam radiation and at an unspecified dose of Al-K_{α} x-ray radiation. The ultimate utility of this "self-developing" resist system will depend upon its efficacy as an etch barrier. It seems clear that such materials would not serve as adequate etch masks for plasma environments, since they are extremely sensitive to short wavelength radiation and high fluxes of such radiation are produced in a plasma discharge.

3.9 Mid UV Resists

The drive towards higher circuit density in microelectronic devices serves as an impetus for continued improvement in the resolution of optical projection printing technologies. One method of achieving an improvement in resolution is through the use of higher energy, shorter wavelength radiation than that currently employed in the near UV (350 to 450 nm) spectral region. Since the diffraction limited resolution of optical projection printing tools is directly proportional to the exposure wavelength, it is clear that a reduction in the exposure wavelength from the near UV (NUV) to the 313 nm, mid UV (MUV) emission line, or to the 254 nm line in the deep UV (DUV) would produce a substantial increase in resolution.

MUV lithography where exposure is based on the 313 nm mercury emission, is a relatively mature technology. The resolution enhancement that accrues from a shift to the MUV region has been carefully documented (*69*) and MUV projection printers are commercially available at this time. These include the Perkin Elmer Micralign series 300 and 500 both of which have MUV capability.

The spectral irradiance of a typical mercury xenon lamp is shown in Figure 45. There is a characteristic high intensity mercury emission at 313 nm and a low intensity emission at 334 nm The MUV projection printing tools isolate these two lines through insertion of band-pass filter sets into the optical path, such that little or no light to the blue of 300 nm or to the red of 350 nm is transmitted to the resist surface. The transmission of a filter set of this sort for the Perkin Elmer Micralign 500 is shown in Figure 46.

Commercially available positive photoresists demonstrate greatly reduced sensitivity in the MUV in comparison to the performance in the NUV. The reasons for this loss of performance with reduced wavelength are severalfold: first, the absorptivity (molar extinction) of the sensitizers used in formulating most commercial positive resists, is very low in the MUV compared to that in the NUV (Figures 19-20). Secondly, these materials undergo photochemistry ultimately leading to a photo-product that is transparent at 405 nm but which absorbs at 313 nm; that is, the sensitizer bleaches cleanly in the NUV but not in the MUV. Thirdly, the phenolic resins used to formulate many of the commercial resists have a significant unbleachable absorbance at 313 nm but are essentially transparent at about 350 nm (Figure 22).

The consequence of these accumulated undesirable optical characteristics has been quantified in several studies (*70-71*). In general, these characteristics result in inefficient light absorption, attenuation of dose with

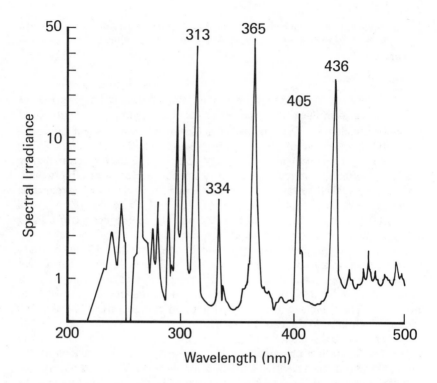

Figure 45. The relative spectral irradiance of a high pressure mercury-xenon lamp of the sort commonly used in optical aligners.

film thickness, and profile degradation. One indication of this is shown in Figure 47 which presents SAMPLE (*15*) generated resist profile simulations. The resist parameters are those of AZ1350J except that the unbleachable absorbance was varied from 0.08 μm^{-1} to 0.43 μm^{-1}. The exposure dose was varied to open the 3 μm pitch grating at constant development time and to dimension. Note the loss in sensitivity and degradation in wall profile that result from increased absorbance.

Though certain positive resist formulations that are commercially available have been shown to function satisfactorily under MUV exposure conditions in terms of the image quality that can be generated (*71,72*), none of these provides a throughput capability comparable to that obtainable under NUV conditions. IBM researchers have recently described the development of a resist designed specifically for exposure in the MUV spectral region (*19*). The resist is a two-component, positive system based on

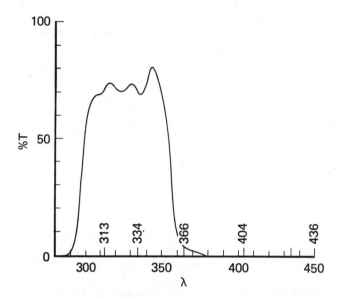

Figure 46. The optical transmission characteristics of a UV-3 filter for the Perkin Elmer Micralign 500. The filter was designed for Mid-UV projection lithography. Note that it is essentially opaque in the near and deep UV. The wave length of the major mercury emission lines are noted.

the use of a new diazonapthoquinone sensitizer and a novolac matrix resin. The novolac resin was chosen to provide optimal transmission characteristics in the MUV, and the structure of the sensitizer was adjusted to provide increased optical absorbance at the 313 nm emission line and designed to undergo photochemistry to produce products that are transparent at this wavelength. Figure 48 shows an unexposed and bleached absorbance spectrum of the IBM-MUV resist cast on quartz. This spectrum should be compared with that of AZ1350J photoresist (Figure 22). Note that AZ1350J bleaches effectively at 405 nm, in the NUV but does not bleach at all at 313 nm in the MUV. The new resist bleaches effectively across the range of wavelengths from 300 to 450 nm. Scanning electron micrographs of 3 μm pitch gratings printed in the new MUV resist are provided in Figure 49.

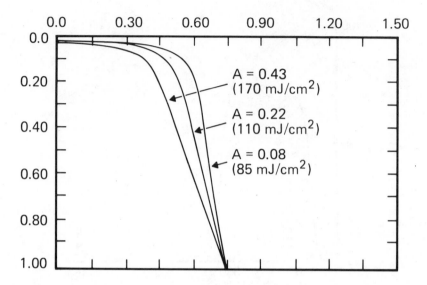

Figure 47. SAMPLE generated profile simulations in AZ1350J resist exposed at 436 nm on an index matched substrate. The mask edge is at 0.75μm on the horizontal scale. The unbleachable optical absorbance (A) was varied from the actual value at 436 nm, 0.08 to 0.43. The dose was adjusted to develop each case to dimension (1.5μm space) at constant development time. As A increases, all other factors being constant, the dose required to open the line increases and the resist profile becomes shallower.

3.10 Deep UV Resists

As described above, solutions to most of the problems of MUV lithography seem to be in hand, both from the resist and exposure tooling point of view. Deep UV lithography (DUV) based on exposure at wavelengths below 300 nm, presents far more difficult technical challenges. Complete solutions to the DUV tooling and resist problems are not yet available, but substantial progress is being made in several laboratories.

The tooling issues association with the change from NUV to DUV lithography fall into two general categories: lens design (and fabrication) and source brightness. The lens problem is a materials issue and is related to the fact that the number of optical-grade materials with acceptable transmission characteristics below 300 nm is limited. In particular, it is

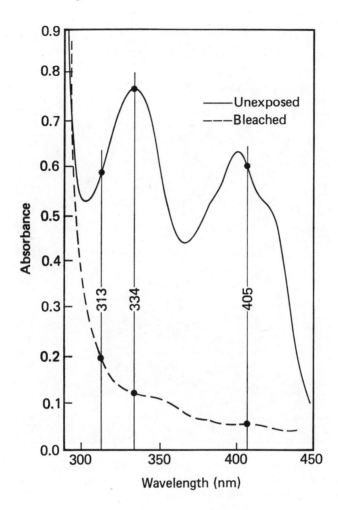

Figure 48. Ultraviolet absorbance spectrum of IBM's experimental mid-UV resist. The resist was coated on quartz and the spectrum was recorded before and after exposure. This figure should be compared to Figure 24 with respect to the extent of bleaching at 313 and 334 nm.

difficult to find transparent materials with the range of refractive index characteristics required to allow chromatic aberration correction to be built into compound lens designs. Reflecting systems do not suffer from this limitation; hence, either fully reflective or compound reflecting/refracting designs for DUV projection tools would seem, at this time, the path of choice.

Figure 49. Scanning electron micrograph of images printed in IBM's experimental Mid-UV resist using a Perkin Elmer Micralign 500 in the UV-3 mode. The resist thickness is 1.1μm the period of the grating is 3μm.

DUV source brightness is a problem for several reasons: First, the intense, allowed mercury emission line at 254 nm, which would be ideal for DUV lithography, undergoes efficient self-absorption with the result that high pressure discharge sources of the sort currently in use in projection tools, produce virtually no output at 254 nm (See Figure 45). Low pressure lamps actually produce most of their radiated power at 254 nm. but are of much lower total brightness than the high power, high pressure lamps that radiate at longer wavelengths. Secondly, a given incident dose in mJ/cm^2

of DUV radiation represents only about half as many photons as the corresponding dose of NUV radiation simply because of the difference in energy per photon which is, of course, proportional to frequency ($E = h\nu$). Consequently, for a resist material with the same quantum efficiency, twice as much dose is required in the DUV as in in the NUV in order to carry out an equal number of photochemical reactions.

Since classical sources have lower output power in the DUV than in the near UV and because what power is available at DUV wavelengths is only half as chemically efficient, new sources and/or very sensitive resist materials must be developed in order to realize the resolution enhancement that DUV lithography offers without suffering intolerably long exposure times. Fortunately, recent experiments with unconventional sources have demonstrated the feasibility of producing a very high flux of DUV radiation that is useful in imaging technology. Major advances have also been made in the design and synthesis of novel high sensitivity DUV resist materials.

Whereas a one kilowatt mercury-xenon lamp may provide a total collected power of 10 to 20 milliwatts in the DUV, excimer lasers can provide 10 to 20 watts of power at any one of several wavelengths in the DUV. Of particular interest are the KrCl and KrF excimer lasers which have outputs at 222 and 249 nm, respectively. High quality contact printed images have been demonstrated using excimer laser radiation (73,74), and early projection printing experiments using excimer laser sources have also been described (75). The laser source is an attractive one for projection printing, since the spatial coherence is such that minimal chromatic aberration correction is required in the lens design; yet, because of relatively low temporal coherence, the excimer lasers do not suffer from speckle. Implementation of excimer laser sources into practical projection tools will require extensive redesign of the condenser optics and demands consideration of the fact that the laser provides a pulsed source rather than the CW output available from conventional mercury lamps.

High intensity, microwave powered emission sources have recently been developed that are reported to provide substantially higher DUV output than classical electrode discharge mercury lamps (76). These sources suffer from self-absorption of the intense 254 nm emission but have a relatively high output in a band between 240 and 280 nm. They are extended sources of finite size rather than point sources, and they must also be an integral part of a tuned, resonant microwave cavity. Consequently, extensive condenser design work would be required in order to utilize the microwave powered sources in projection printers.

The earliest work in DUV lithography utilized PMMA resist (77) and required exposure times of tens of minutes. Researchers were forced to tolerate the excessive exposure times because conventional, positive NUV resists are not useful for DUV exposure. These NUV resist materials (the diazoquinone-novolac systems) suffer from the same two factors that limit their utility in the MUV, namely unbleachable sensitizer and photoproduct absorbance and strong absorption of the novolac resin. These issues only limit utility in the MUV but they are sufficiently severe as to preclude utility in the DUV. See, for example, Figures 22, 23 and 48.

Two interesting attempts to redesign these resist materials have been reported. The first consisted of altering the structure of the sensitizer such that it bleaches in the DUV (70). The resulting resist provided adequate sensitivity but suffered from sensitizer volatility and solubility problems and profile degradation was experienced in films over 0.5 μm micron thickness due to the unbleachable absorbance of the matrix resin from which the resist was formulated.

Wilkins and coworkers have redesigned both the sensitizer and the matrix resin (78-79). They have tested a variety of o-nitrobenzyl esters of cholic acid as sensitizers. These substances, like the diazoquinones, are insoluble in aqueous base but undergo a photo-reaction that yields base soluble products. The matrix resin chosen for the new sensitizer materials is a copolymer of methyl methacrylate and methacrylic acid that is far more transparent than novolac resins in the DUV. The new resist materials are reported to have useful sensitivity (ca. $100 mJ/cm^2$) and extremely high contrast. The resist formulation is essentially aliphatic in nature and would be expected to be less stable to dry etching environments than the aromatic-based novolac resin materials (24).

Attempts to improve the DUV sensitivity of PMMA have spawned a variety of new DUV resist materials. Notable among these are copolymers of methyl methacrylate and indenone (80) which are reported to provide positive-tone resist function at 20 to 60 mJ/cm^2 in the DUV and copolymers of methyl methacrylate and 3-oximino-2-butanone (81). The latter materials provide a substantial increase in sensitivity over PMMA and are capable of 1 micron resolution.

Another interesting positive-tone polyacrylate DUV resist has been reported by Ohno and coworkers (82). This material is a copolymer of methyl methacrylate and glycidyl methacrylate. Such materials are negative e-beam resists, yet in the DUV they function as positive resists. Thermal crosslinking of the images after development provides relief structures with exceptional thermal stability. The reported sensitivity of these copolymers is surprising, since there are no obvious scission mechanisms available to the system other than those operative in PMMA homopolymer, and the glylcidy side-chain does not increase the optical density of the system.

Negative DUV resist systems have also been developed, some of which have demonstrated extremely high sensitivity. These include poly(butyl α-chloroacrylate) (83) which is reported to be at least 500 times as sensitive as PMMA and partially chlorinated, narrow dispersity poly(p-vinyl toluene) (84), which has been imaged at $10 mJ/cm^2$ in contact printing. Both of these resists suffer some resolution limitation due to swelling.

A novel, negative DUV resist has been developed that is devoid of swelling phenomena. This material is a bis-arylazide sensitized poly(vinylphenol) (85). The material is strongly absorbing in the DUV. Dissolution inhibition occurs upon exposure, but mainly in the upper layers of the film. Consequently, resist profiles change during development from overcut to vertical to undercut with increasing development time. High quality DUV projection printed images that show no evidence of distortion due to swelling have been made at high throughput with this new resist. The material does not swell and because of its high absorbance does not suffer image distortion due to reflected light interference phenomena. The major limitation of the system seems to be the narrow process window. Image profile and linewidth change rapidly with development time. Consequently, reproducibility requires extremely careful control of development and process conditions.

Ito and coworkers (86-87) recently described a new DUV resist system that has high sensitivity, does not suffer swelling during development, and may be used either as a positive or negative system. The new resist is based on a formulation consisting of t-butyloxycarbonyl (t-BOC) protected poly(vinylphenol) 88 (Figure 50) and an onium salt sensitizer. Exposure

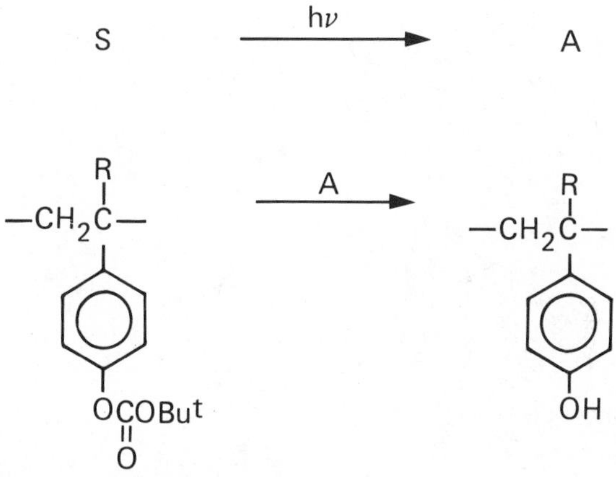

Figure 50. Schematic function of the t-Bocstyrene resist. S is a sensitizer such as diphenyliodonium hexafluoroarsenate which undergoes radiolysis to produce a strong acid (A). The acid attaches the side chain of the poly-(t-Bocstyrene) where it catalyzes acidolysis of the carbonate to liberate CO_2 and isobutylene and free the phenolic hydroxyl group to produce poly(p-hydroxystyrene) in the exposed areas of the resist film. The acid A is a catalyst and can cleave many carbonate groups.

results in photolysis of the salt to produce a local concentration of a strong acid. The acid reacts with the polymer side-chain where it acts as a catalyst for acidolysis of the t-BOC group, resulting in liberation of carbon dioxide, isobutylene, and free poly(vinylphenol). Since acid is not consumed in the deprotection reaction, one mole of photo-generated acid may cleave many t-BOC groups, thus providing high sensitivity through chemical amplification.

Development of the exposed resist in nonpolar solvents selectively removes unreacted resist to give a negative tone image. Development with a polar solvent (or aqueous base) selectively dissolves the exposed (phenolic) area and provides positive tone images. The sensitizers described are the same as those used in the poly(phthalaldehyde), self-developing resist. Scanning electron micrographs of images printed in this new resist system are shown in as Figure 51. The extent to which inclusion of sensitizer materials like hexafluoroarsenate salts in the resist will have untoward effects on device performance is not yet clear.

Developments in DUV resist materials and tooling promise another generation of optical lithography for device manufacturing. The ultimate resolution that will be obtainable by DUV lithography in a production environment will depend on continued developments in exposure tool design, alignment systems, and resist technology but the potential for submicron manufacturing by DUV lithography appears extremely promising.

Figure 51. Scanning electron micrographs of images printed in the poly (t-Bocstyrene) resist. The images were generated by exposing a wafer coated with resist, splitting the wafer in half, then developing one fragment in a polar solvent to obtain a positive tone image (left) and the other half in a nonpolar solvent to obtain a negative tone image (right).

Literature Cited

1. DeForest, W. "Photoresist Materials and Processes," McGraw Hill, New York, New York, 1975, p 11
2. Calvert, J. G.; Pitts, J. N., Jr. "Photochemistry" John Wiley and Sons, Inc., New York, New York, 1967.
3. Guillet, J. E.; Houvenaghel-Defoort, B.; Kilp, T.; Turro, N. J.; Steinmetzer, H. C.; Shuster, G. *Macromolecules*, 1974, *7* (6), p 942.
4. Amerik, Y.; Guillet, J. E. *Macromolecules*, 1971, *4* (4), p 375.
5. Charlesby, A. "Atomic Radiation and Polymers," Pergamon Press, Oxford, 1960.
6. Chapiro, A. "Radiation Chemistry of Polymeric Systems," Interscience, New York, 1962.
7. Dole, M. "The Radiation Chemistry of Macromolecules," Vol. 1, Academic Press, New York, 1972.
8. O'Donnell, J. H.; Rahman, N. P.; Smith, C. A.; and Winzor, D. J. *Macromolecules*, 1979, *12* (1), p 113 and references therein.
9. Pittman, C. U., Jr.; Iqbal, M.; and Chen, C. Y. *J. Polymer Sci., Polymer Chem. Ed.*, 1978, *16* (10), p 2721.
10. Hatzakis, U.; Ting, C.; Viswanathan, N. "Fundamental Aspects of E-Beam Exposure of Polymer Resists," Electron Ion Beam Science & Technology, Sixth International Conference, San Francisco, May 1974.
11. Alexander, P.; Black, R. M.; Chorlesby, A. *Proc. Royal Soc.* (London), 1955, *A232*, p 31.
12. Alexander, P.; Charlesby, A.; Ross, M, *Proc. Royal Soc.* (London), 1954, *A223*, p 392.
13. Kilb, R. W. *J. Phys. Chem.*, 1959, *63*, p 1838.
14. Kyser, D. F.; Pyle, R. *IBM J. Res. and Dev.*, 1980, *24*, p 426.
15. Oldhom, W. G.; Nandgaonkor, S. N.; Neureuther, A. R.; O'Toole, M. *IEEE Trans. Electron Devices*, 1979, *ED-26*, p 717.
16. Konnerth, K. L.; Dill, F. H. *IEEE Trans. on Electron Devices*, 1979, *ED-22*, p 452.
17. Chapter 2 section 2.1.
18. Knop, A.; Scheib, W. "Chemistry and Applications of Phenolic Resins," Springer-Verlag, Berlin, Heidelberg, New York, 1979.
19. Willson, G.; Miller, R.; McKean, D.; Clecak, N.; Tompkins, T.; Hofer, D. "Design of a Positive Resist for Projection Lithography in the Mid UV," Proceedings of the SPE Regional Technical Conference, Ellenville, New York, Nov. 1982, p 111.
20. Pacansky, J,; Lyerla, J. R. *IBM J. Res. Develop.*, 1979, *23*, p 42.
21. Moritz, H.; Paal, G. U.S. Patent No. 4, 104, 070, 1978; *Chem. Abstracts*, 1978, *88*, 14344u.

22. Takahashi, Y.; Shinozaki, F.; Ikeda, T. *Jpn. Kokai Tokkyo Koho,* 1980, *008,* p 8032; *Chem. Abstracts,* 1980, *93,* 195530y.
23. MacDonald, S. A.; Miller, R. D.; Willson, C. G.; Feinberg, G. M.; Gleason, R. T.; Halverson, R. M.; MacIntyre, M. W.; Motsiff, W. T. "Image Reversal: The Production of a Negative Image in a Positive Photoresist," Kodak Microelectronics Seminar, San Diego, 1982.
24. Pederson, L. A. *J. Electrochem. Soc.,* 1982, *129* (1), p 205.
25. Shultz, A. R. *J. Polymer Sci.,* 1959, *35,* p 369.
26. Hiraoka, H. *IBM J. Res. Develop.,* 1977, *21,* p 121, and references therein.
27. Campbell, D. *J. Poly. Sci.,* 1970, *4,* p 97.
28. Tsuji, K. *Adv. Poly. Sci.,* 1973, *12,* p 131.
29. Hiraoka, H. *Macromolecules,* 1976, *9,* p 359.
30. Hiraoka, H. *J. Amer. Chem. Soc.,* 1973, *95,* p 1664.
31. Moreau, W.; Merrit, D.; Moyer, W.; Hatzakis, M.; Johnson, D.; Pederson, L. *J. Vac. Sci. Technol.,* 1979, *16* (6), p 1989.
32. There are a large number of such studies. For leading references see the work of C. U. Pittman, Jr., J. N. Helbert, J. H. Lai, C. F. Cook, M. Hatzakis, S. Nonogaki, H. Ito, etc.
33. FBM Resist is Marketed by Daiken Kogyo of Japan.
34. Harada, K. *J. Electrochem. Soc.,* 1980, *127,* p 491.
35. Tada, T. *J. Electrochem. Soc.,* 1979, *126,* p 1831.
36. Solonia, W. *J. Russ. Phys. Chem. Soc.,* 1898, *30,* p 826.
37. Fredernick, D. S.; Cogan, H. D.; Marvel, C. S. *J. Amer. Chem. Soc.,* 1934, *56,* p 1815.
38. Brown, J. R.; O'Donnell, J. H. *Macromolecules,* 1970, *3,* p 265.
39. Brown, J. R.; O'Donnell, J. H. *Macromolecules,* 1972, *5,* p 109.
40. Thomspon, L. F.; Bowden, M. J. *J. Electrochem. Soc.,* 1973, *120,* p 1722.
41. Bowmer, T. N.; O'Donnell, J. H. *Radiat. Phys. Chem.,* 1981, *17,* p 177 and references therein.
42. Bowden, M. J.; Thompson, L. F.; Farenholtz, S. R.; Doerries, E. M. *J. Electrochem. Soc.,* 1981, *128,* p 1304.
43. Thompson, L. F.; Feit, E. D.; Heidenreich, R. D. *Polymer Engin. and Sci.,* 1974, *14* (7), p 529.
44. Thompson, L. F.; Ballantyne, J. P.; Feit, E. D. *J. Vac. Sci. Technol.,* 1975, *12* (6), p 1280.
45. Sogo, K.; Tanaka, Y.; Uchiho, K. "Fabrication of High Quality E-Beam Master Masks," Kodak Microelectronics Seminar, Interface '80, October, 1980.
46. Ohnishi, Y.; Itoh, M.; Mizuko, K.; Gorkan, H.; Fujiwara, S. *J. Vac. Sci. Technol.,* 1981, *19* (4), p 1141.

47. Shiraishi, H.; Taniguchi, Y.; Horigomi, S.; Nonogaki, S. *Polymer Engineering and Sci.,* 1980, *20* (16), p 1054.
48. Feit, E. D.; Thompson, L. F.; Wilkins, C. W., Jr.; Wurtz, M. E.; Doerries, E. M.; Stillwagon, L. E. *J. Vac. Sci. Technol.,* 1979, *16* (6), p 1997.
49. Feit, E.; Stillwagon, L. *Polymer Engineering and Sci.,* 1980, *20* (16), p 1058.
50. Liutkis, J.; Parasczak, J.; Shaw, J.; Hatzaski, M. "Poly-4-Chlorostyrene, A New High Contrast Negative E-Beam Resist," SPE Regional Technical Conference, Ellenville, New York, Nov. 1982, p 223.
51. Choony, H. S.; Kohn, F. J. *J. Vac. Sci. Technol.,* 1981, *19* (4), p 1121.
52. Feit, E. D.; Wurtz, M. E.; Kammlott, G. W. *J. Vac. Sci. Technol.,* 1978, *15*, p 944.
53. Gong, B. M.; Yee, Y. D.; Gu, Q. M.; Zhang, Q. B. *J. Vac. Sci. Technol.,* 1979, *16* (6), p 1980.
54. Crank, J.; Park, G. S. "Diffusion in Polymers," Academic Press, New York, 1968. Chapter 7 by K. Ueberreiter, "The Solution Process," p 219.
55. Hofer, D. C.; Kaufman, F. B.; Kramer, S. R. Aviram, A. *Appl. Phys. Lett.,* 1980, *37* (3), p 314.
56. Murase, K.; Kakuchi, M.; Sugawara, S. "Les Films Sensikles anx Electrons et aux Rayons X," International Conference on Microlithography, Paris, June 1977.
57. Brault, R. G.; Miller, L. J. *Polymer Engin. and Sci.,* 1980, *20* (16), p 1064.
58. Taylor, G. N.; Wolf, T. M. *J. Electrochem. Soc.,* 1980, *127*, p 2665.
59. Taylor, G. N. *Solid State Technology,* 1980, *23* (5), p 73
60. Bowden, M. J.; Thompson, L. F. *Polymer Engin. Sci.,* 1974, *14*, 525.
61. Smith, J. N.; Hughs, H. G.; Keller, J. V.; Goodner, W. R.; Wood, T. E. *Semiconductor International,* 1979, p 41.
62. Taylor, G. N.; Wolf, T. M. *J. Electro Chem. Soc.,* 1980, *127*, p 2665.
63. Taylor, G. N. *Solid State Technology,* 1980, *2* (9), p 73.
64. Tsuda, M.; Oikawa, S.; Kanai, W.; Yokota, A.; Hijikata, I.; Uehara, A.; Nakane, H. *J. Vac. Sci. Technol.,* 1981, *19*, p 259.
65. Yokota, A.; Yabuta M.; Kanai, W.; Kakhiwaga, K.; Hijikata, I.; Nakane, H. "Plasma Developable Photoresist Containing Electronic Excitation Energy Quenching System," SPE Regional Technical Conference, Ellenville, New York, Nov. 1982.
66. Willson, C. G.; Ito, H.; Frechet, J. M. J.; Houlihan, F. *Proceedings of IUPAC 28th Macromolecular Symposium,* Amherst, Massachusetts, July 1982, p 448.

67. Ito, H.; Willson, G., "Chemical Amplification in the Design of Dry Developing Resist Materials," SPE Regional Technical Conference, Ellenville, New York, Nov. 1982.
68. Aso, C.; Tagomi, S.; Kunitake, T. *J. Polym. Sci.*, 1969, 7, (A-1), p 497.
69. Bruning, J. H. *J. Vac. Sci. and Technol.*, 1979, *16*, p 1925.
70. Grant, B. D.; Clecak, N. J.; Twieg, R. G.; Willson, C. G. *IEEE Trans. Electron. Devices,* 1981, *ED28* (11), p 1300.
71. Hofer, D. C.; Willson, C. G.; Neureuther, A. R.; Hahey, M. *Proceedings of SPIE*, 1982, *334*, p. 196.
72. Robic, J. P.; Knight, S.; Straub, W. "Mid-UV Perkin Elmer 500 Lithography with AZ4110 Resist: A Multifactorial Tool and Process Evaluation," International Conference on Microlithography, Grenoble, France, Oct. 1982.
73. Jain, K.; Willson, C. G.; Liu, B. J. *IEEE Electron Device Letters,* 1982, *EDL-3* (3), p 53.
74. Jain, K.; Willson, C. G.; Liu, B. J. *IBM J. Res. and Develop.*, 1982, *26* (2), p 151.
75. Dubroeucz, G. M.; Zahorsky, D. "KrF Excimer Laser as a Future Deep UV Source for Projection Printing," International Conference on Microlithography, Grenoble, France, Oct. 1982.
76. Ury, M. G.; Mathews, J. C.; Wood, C. H. *Proceedings of SPIE*, 1982, *334*, p 241.
77. Moreau, W. M.; Schmidt, P. R.   1038th Electrochemical Society Meeting, Extended Abstract #187, 1970.
78. Reichmanis, F.; Wilkins, C. W., Jr.; Chandross, E. A. *J. Vac. Sci. Technol., 1981, 19*, p 1338.
79. Wilkins, C. W., Jr.; Reichmanis, E.; Chandross, E. A. "Lithographic Evaluation of an o-Nitrobenzyl Ester Based DUV Resist System," *J. Electrochem. Soc.*, in press.
80. Hartless, R. L.; Chandross, E. A. *J. Vac. Sci. Technol*, 1981, *19*, p 1333.
81. Wilkins, C. W., Jr.; Reichmanis, E.; Chandross, E. A. *J. Electrochem. Soc.*, 1980, *127*, p 2510 and p 2514.
82. Yamashita, Y.; Ogura, K.; Kunishi, M.; Kawazu, R.; Ohno, S.; Mizokami, Y. *J. Vac. Sci. Technol.*, 1979, *16*, p 2026.
83. Japan Economic Journal, Sept. 7, 1982.
84. Harita, Yoshiyuki, Personal Communication. To be published.
85. Iwayanagi, T.; Kohashi, T.; Nonogaki, S.; Matsuzawa, T.; Douta, K.; Yanazawa, H. *IEEE Trans. Electron Devices*, 1981, *ED-28*, p 1306.
86. Frechet, J. M. J.; Ito, H.; Willson, C. G. "Sensitive Deep UV Resist Incorporating Chemical Amplification," International Conference on Microlithography, Grenoble, France, Oct. 1982.

87. Ito, H.; Willson, C. G.; Frechet, J. M. J. "A Sensitive Deep UV Resist System," SPE Regional Technical Conference, Ellenville, New York, Nov. 1982.

88. Frechet, J. M. J.; Eichler, E.; Ito, H.; Willson, C. G. "Poly (p-t-Butyloxycarbonyloxystyrene): A convenient precursor to p-hydroxystyrnene resins," Accepted for Publication in *Polymer*, 1982.

RECEIVED February 10, 1983

4

Resist Processing

L. F. THOMPSON and M. J. BOWDEN

Bell Laboratories, Murray Hill, NJ 07974

0097-6156/83/0219-0161$14.25/0
© 1983 American Chemical Society

4.1 Introduction

There are five distinct stages or phases involved in the development of a viable resist system that collectively may involve many years of effort in order to successfully introduce the resist into a manufacturing environment. These stages are as follows:

1. Fundamental research on polymer and radiation chemistry

2. Synthesis and characterization of potential materials

3. Initial lithographic evaluation

4. The engineering scaleup of the synthesis of a specific, identified material

5. Process development for manufacturing application

The first four stages are well founded in terms of engineering concepts and fundamental scientific principles. Knowledge of the chemistry, physics and chemical engineering of resist design and synthesis allows these steps in the resist development sequence to be performed with a good degree of understanding and purpose. By comparison, the fifth step, process development, is largely empirical and is by far the most difficult, least understood, and time consuming of all of the project phases, requiring some 5 to 10 times more effort than the first four steps. Indeed, it is our feeling that many good materials have been identified and subsequently rejected because insufficient process engineering was done.

Much of the time spent on process development is in optimizing the wide range of variables that must be examined if one is to obtain the requisite lithographic pattern quality. Unlike the first four phases, the methodology required, and the strategy used in process development are largely an art and not a science. An example of this can be seen by tracing the history of chromium mask fabrication from its infancy some 15 years ago to its present state. During this evolutionary process, many serious processing problems were encountered which were given such entertaining names as mouse nips, moose nips, dog tails, chicken feet etc. It is obvious from this "animal farm" description that little was understood initially about any one of these problems, although each was eventually solved. This chapter represents advances made in understanding the art of processing; however, much work remains before processing can be considered a science.

The goal of any lithographic technology is to produce a three-dimensional relief image whose size and fidelity match, as closely as possible, that of the image in the mask or the serial exposure beam. As we discussed in Chapter 2, pattern definition in a resist consists of two distinct steps:

1. The formation of a latent image in the resist by exposure to radiation

2. The development of that image to give a three-dimensional relief structure suitable for subsequent pattern transfer

The quality or fidelity of the latent image is governed by the physics and chemistry of the exposure step as discussed in Chapters 2 and 3. This chapter deals with the processing steps (outlined in Figure 1) that are involved in the development of the latent image into the relief image. We will first examine the important performance criteria such as sensitivity, resolution, linewidth control and defect density by which processing is evaluated. Each processing step will then be described in detail, together with its relevance to these performance criteria. Where appropriate, scientific principles that can be used as a guide in optimizing each step will be introduced. As we will see later, there are numerous variables associated with each processing step, many of which are interactive with subsequent steps, making optimization an extremely time-consuming process. It is the author's intent for this chapter to serve as a general guide to understanding the various processing steps used for all polymeric resist systems. There are specific considerations for some resists and these will be discussed as appropriate.

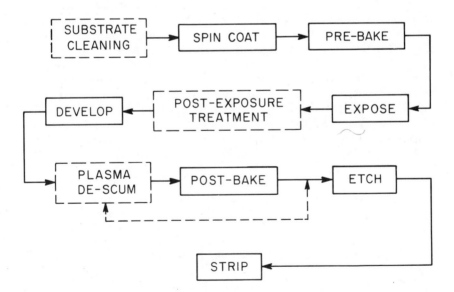

Figure 1. Flow chart for a typical resist process. Steps in broken lines are not used for all materials.

4.2 Performance Criteria

The capabilities (or performance standards) of any lithographic technology are evaluated with respect to five major criteria:

1) Sensitivity and contrast

2) Resolution

3) Linewidth control

4) Defect density

5) Etch resistance

In addition, there are four other resist-related criteria, viz., adhesion, shelf life, supply and quality assurance that must be taken into account. Lithographic performance with respect to these parameters is dictated and/or limited by the *hardware*, *materials*, and *processing* used to generate the requisite pattern as summarized in Table I.

4.2.a Sensitivity and Contrast. The sensitivity and contrast of a resist are determined primarily by the chemical composition, molecular parameters, and physical properties of the polymer as discussed in Sections 3.3 and 3.4 and summarized in Table II. In order not to limit throughput it is mandatory that the resist exhibit a sensitivity commensurate with the exposure parameters of the machine. As a result, hardware operating parameters determine the sensitivity requirements of the resist. These are summarized in Table III for each lithographic technology. It is important to note in this regard that there are practical limits to sensitivity. For example, an electron beam resist with a sensitivity greater than $\sim 10^{-8}$ C cm^{-2} would undergo thermal reactions at room temperature and would be unsatisfactory because of unacceptably short shelf life of both the resist solution and spun films ([1,2]). The lower limit of sensitivity is governed by throughput considerations. Figure 2 illustrates the sensitivity "window" for electron resists and similar limits can be defined for other exposure technologies. Clearly, processing must provide us with the sensitivity and contrast of which the resist is intrinsically capable. The developing criteria and conditions can affect the sensitivity by as much as an order of magnitude and must be optimized so as to yield not only the highest sensitivity, but also the desired resolution (Section 4.2.b), linewidth control (Section 4.2.c) and defect density (Section 4.2.d).

Sensitivity is conventionally defined as the incident input energy (or dose) per unit area required to achieve the desired chemical response in the resist, i.e.,

$$D = \frac{E}{A} \tag{1}$$

TABLE I. Parameters Affecting Lithographic Performance

Performance Criteria	Hardware Parameters[1]	Process Parameters	Material Parameters
SENSITIVITY	Energy density Exposure rate Exposure environment	Developing Post-exposure treatment	Radiation efficiency Molecular weight
RESOLUTION	Scattering Address size Mechanical stability Mask quality	Developing Baking Descum Etching	Contrast Swelling T_g Molecular weight distribution
ETCH RESISTANCE	-	Type of etch Baking	Composition Stability T_g
LINEWIDTH CONTROL	Lens imperfections Source stability Mask dimensions Beam shape Mechanical stability	Developing Etching Baking Plasma descum Film thickness Film uniformity	Swelling
DEFECTS	Mask defects Hardware contamination	Cleanliness Number of process steps	Filtering Gelling
ADHESION	-	Developing Baking	Composition
SHELF LIFE	-	-	Reactivity (sensitivity) Storage conditions

[1] Depends on type of exposure system.

TABLE II. Polymer and Processing Parameters
that Affect Sensitivity and Contrast of a Resist

	PARAMETER
POLYMER PROPERTIES	Chemical composition Molecular weight Molecular weight distribution Glass transition temperature (T_g) Density Average atomic number[1]
PROCESSING	Developing criteria Developing conditions Developer composition Pre and Postbaking conditions

[1]For X-ray; E-beam and Ion beam.

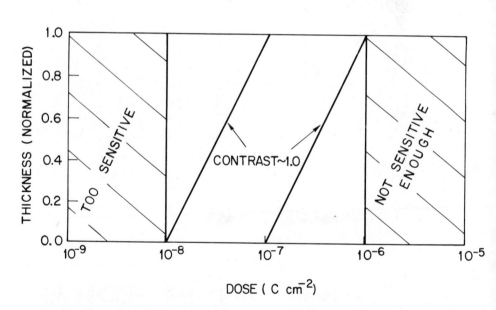

Figure 2. A sensitivity diagram showing the exposure "window" for electron resists.

TABLE III. Hardware Parameters that Determine the
Sensitivity and Contrast Requirements of a Resist.

Lithography	Typical Units of Sensitivity	Relevant Parameters (other than energy flux)
PHOTO-	mJ cm^{-2} μJ cm^{-2} J cm^{-2}	Wavelength (spectrum) Substrate reflectivity Mask contrast Exposure ambient Source Brightness Flash time
E-BEAM	C cm^{-2} μC cm^{-2}	Accelerating voltage Substrate backscatter Total time in vacuum[1] Beam size and shape Source brightness Flash (dwell) time
X-RAY	mJ cm^{2} μJ cm^{-2}	Wavelength Exposure ambient Mask absorption characteristics Source brightness Exposure time
ION BEAM	Particles cm^{-2} eV cm^{-2}	Ion mass and charge Accelerating voltage Beam size Beam shape Flash (dwell) time

[1] only for materials that exhibit a post exposure reaction.

where D is the dose, E the incident energy in appropriate units, and A the exposed area, usually expressed in square centimeters (*3,4*). Table III lists some commonly used units of sensitivity for the various lithographic strategies. The *lithographically useful sensitivity* has been defined by Ballantyne (*5*) as the dose per unit area that results in dimensional equality of clear and opaque features which are nominally equal in pattern design. Figure 3 shows the variation in feature size with dose for a typical negative electron resist (*6*). The incident dose required to produce the requisite lithographic pattern is referred to as sensitivity, (D_g^x in Figure 3).

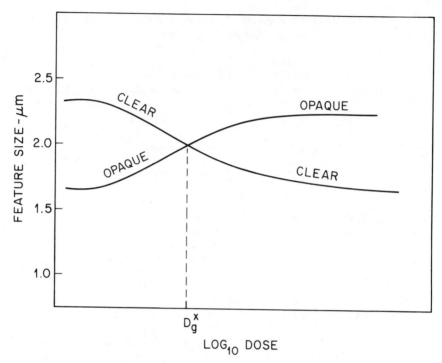

Figure 3. Size of clear and opaque 2.0 µm features as a function of exposure dose for a negative electron beam resist. The dose (D_g^x) that results in the correct feature size is denoted as the "sensitivity".

The resolution capability of a resist is directly related to resist contrast (γ) which, for a negative resist, is related to the *rate* of crosslinked network formation at a constant input dose. It is somewhat more complicated for a positive resist being related to the rate of chain scission and the rate of change of solubility with molecular weight with the latter being markedly solvent dependent. Contrast, like sensitivity, is governed by the type of chemical reactions that occur in the polymeric resist and is affected by molecular parameters such as molecular weight distribution and chemical composition.

Sensitivity and contrast are conveniently measured experimentally by exposing areas of resist of known size to varying radiation doses and measuring the film thickness remaining after development for each area. In the case of negative resists, gel is not formed until a critical dose, denoted as the interface gel dose (D_g^i), has been reached. At this dose no lithographically useful image has been formed since the film thickness is insufficient to serve as an etching mask. Thereafter, the gel content (film thickness) increases with increasing dose until the thickness remaining is equal to the original film thickness. (In practice, this maximum thickness is slightly less than the original film thickness due to volume contraction during crosslinking.) A plot is then generated of normalized film thickness as a function of log dose as shown in Figure 4 from which the contrast is obtained as

$$\gamma_n = 1/(\log D_g^o - \log D_g^i) = \left[\log \frac{D_g^o}{D_g^i} \right]^{-1} \tag{2}$$

where D_g^0 is the dose required to produce 100% initial film thickness and is determined by extrapolating the linear portion of the normalized thickness vs. dose plot to a value of 1.0 normalized film thickness (7). The sensitivity is taken as the dose (near $D_g^{0.5}$) which crosslinks the film to the required thickness after development. This dose (D_g^x) is determined from Figure 3 as the dose where the size of opaque and clear features are equal.

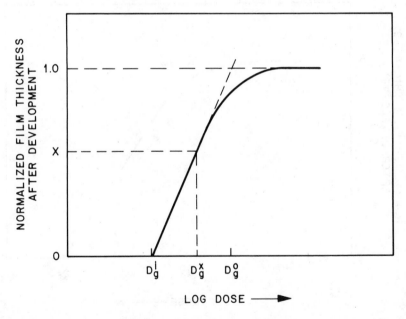

Figure 4. Typical response or sensitivity curve for a negative electron resist. The value of D_g^x is obtained from Figure 3 and usually occurs at 0.5 - 0.7 normalized thickness.

For a positive resist, the film thickness of the irradiated region after development decreases until eventually a critical dose D_P is reached which results in complete removal of the film (8,9). The sensitivity and contrast (γ_p) are evaluated in a manner similar to that for a negative resist. After they have been spin-coated and prebaked, a series of pads of known area are exposed to varying doses. The substrate is developed in a solvent that does not attack the unexposed film and the thickness of the film remaining in the *exposed* areas measured. The film thickness is normalized to the original thickness, and this value is plotted as function of log dose, as shown in Figure 5 where D_p represents the sensitivity of the positive resist. Contrast (γ_p) is determined from the extrapolated slope of the linear portion of the response curve as

$$\gamma_p = 1/(\log D_p - \log D_p^o) = \left[\log \frac{D_p}{D_p^o}\right]^{-1} \qquad (3)$$

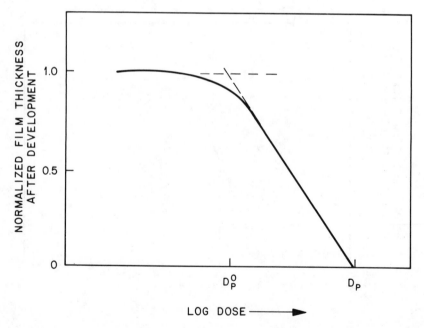

Figure 5. Typical sensitivity or response curve for a positive electron resist. Positive resist response may also be plotted using other conventions (see Chapter 3).

where D_p^o is the dose at which the developer first begins to attack the irradiated film and, as before, is determined by extrapolating the linear portion of the film thickness remaining vs. dose curve. An alternate method of evaluating sensitivity of positive resists has been designed by workers at IBM, San Jose laboratories and is discussed in detail in Section 3.4.a The dose (sensitivity) D_s derived from this method is approximately equal to D_p; however, no comparison can be made of contrast, γ_p as defined in Equation 3.

We have seen that the lithographically useful sensitivity for a positive resist (D_p) is defined as the dose required to produce complete solubility in the *exposed* region, while not affecting the unexposed resist. It is important to note, however, that it is possible to develop a relief image to the substrate for doses less than D_p. This occurs in "force" developing where we allow the developer to attack or *thin* the original, unexposed film (*10-13*) and may give rise to several problems, including loss of linewidth control and possible pinhole formation as a result of thin resist in unexposed areas.

When reporting the sensitivity of a resist, all of the parameters used in the processing should be stated and an exposure curve given. Unfortunately such detail is often omitted in the literature, making comparison of resist sensitivities difficult if not impossible.

4.2.b Resolution. The resolution *capabilities* are determined by hardware, material and processing considerations. The physical and chemical limitations on resolution associated with hardware and materials were discussed in Chapters 2 and 3. Hardware limitations include 1) radiation scattering (diffraction in the case of light), 2) minimum address size or mask features, 3) lens aberrations, and 4) mechanical stability of the system. The most significant factor is scattering which results in the deposition of energy in regions outside the intended patterned area. As mentioned earlier, the effects of radiation scattering can be minimized by using resists that exhibit high contrast.

The primary process-related resist variables associated with resolution include swelling (deformation during development) and stability during etching and baking. These determine or at least influence the choice of developing and baking conditions as well as subsequent pattern transfer techniques. For example, perfect high resolution resist patterns can be destroyed during etching and great care must be exercised in selecting a specific etching process compatible with the resist and the desired film to be etched. All of these factors will be discussed in detail under the various processing steps.

Linewidth (resolution) is usually measured using either transmitted or reflected light. Early optical techniques were manual and used filar (for reflected light) or image shearing eyepieces (transmitted or reflected

light) (14-16). These techniques generally yield a precision of $\pm$ 0.3 μm
(2σ) if at least five measurements per feature are taken. The precision of
the measurement technique is given by

$$\sigma = (\sum_{i=1}^{n} (x_i - \bar{X})^2 / n)^{1/2} \qquad (4)$$

where σ is the standard deviation of a set of n measurements and $\bar{X}$ the
average of all measured values. These methods are time consuming and
operator dependent; as minimum features go below 2.0 μm, linewidth measurement precision must be improved to $\pm$ 0.2 μm (2σ). Automatic optical
linewidth measurement techniques have been available since the late 1970's
and provide a precision of $\pm$ 0.1 μm (2σ) for features 2.0 μm and larger.
Optical techniques are unsatisfactory for features with dimensions less than
1.0 μm where a precision of $\pm$ 0.1 μm is required. Scanning electron
microscope (SEM) techniques may be usually used in these cases. SEM
measurements are taken manually and are very time consuming and care
must be exercised to insure the magnification (both x and y) of the SEM is
calibrated. It is likely that automatic, computer-controlled electron beam
measurement devices will be commercially available in a few years.

An additional complication is that resolution for both positive and
negative tones is affected by the specific pattern geometry. This is known as
the proximity effect. For example, an isolated single line is most easily
resolved in a negative resist, whereas an isolated hole or trench is most
easily defined in positive working materials. Clearly, the resolution capabilities of a resist should always be evaluated and reported for the most difficult
feature to define, since with any integrated circuit pattern, all types of
features will be encountered on a single device. In addition, one should
report all relevant patterning parameters such as accelerating voltage, exposing wavelength, processing conditions, film thickness, and measurement techniques. Without this type of detailed information, it is impossible to compare one material or lithographic system to another. Table IV presents the
factors that affect resolution for each lithographic strategy along with preferred units of measurement.

4.2.c Linewidth Control. This parameter refers to the necessity of maintaining the correct features *size* across an entire substrate and from one substrate to another. This is important since the successful performance of
most devices depends upon control of the *size* of critical structures, as for
example in the gate electrode structure in an MOS device. As feature size
is decreased and circuit elements packed closer together, the margin of error
on feature size control is reduced. The allowable size variation on structures
is generally a fixed fraction of the nominal feature size. A rule of thumb is
that the dimensions must be controlled to tolerances of at least $\pm$ 1/5 the
minimum feature size. Linewidth control is affected by a variety of parame-

TABLE IV. Factors that Affect the Resolution of
Resists for Various Lithographic Options

Lithography	Type of Limitation	Parameter	Preferred Units
PHOTO	Hardware	Wavelength	nm
		Mask feature size	μm
		Coherency (σ)	*
		exposure	
		Ambient	*
		Mask optical density	%T
	Materials	Resist composition	*
		Resist thickness	μm
		Resist T_g	°C
		Substrate	*
		Resist contrast (γ)	
		Molecular weight	*
	Process	Developer and conditions	*
		Pre and Postbaking temperatures	°C
		Pre and Postbaking time	min.
		Plasma descum	*
		Etching technique	*
E-BEAM	Hardware	Accelerating voltage (V)	
		Beam size (shape)	μm
		Pattern geometry	
	Materials	Resist composition	*
		Resist thickness	μm
		Resist T_g	°C
		Resist contrast (γ)	
		Substrate atomic number (Z)	AMU
		Substrate thickness	μm
	Process	Developer	*
		Pre and Postbake temperature	°C
		Pre and Postbake time	min.
		Plasma descum	*
		Etching techniques	*

TABLE IV - Cont.

Lithography	Type of Limitation	Parameter	Preferred Units
X-RAY	Hardware	Wavelength	Å
		Mask structure	*
		Mask contrast	*
	Materials	Resist composition	*
		Resist thickness	μm
		Resist T_g	°C
		Resist contrast (γ)	
		Substrate atomic number	AMU
		Resist absorption	ϵ
	Process	Developer and conditions	*
		Pre and Postbake temperature	°C
		Pre and Postbake time	min.
		Plasma descum	*
		Etching technique	
ION BEAM	Hardware	Same as e-beam	
		Ion mass	AMU
		Ion charge	*
	Materials	Same as x-ray and e-beam	
	Process	Same as e-beam.	

*units or data as appropriate

ters associated with hardware, processes and materials (see Table I) and is generally evaluated by measuring a series of features with known sizes across a substrate and then plotting the feature dimension as a function of *position* on the substrate. The standard deviation at the one or two sigma level is then adopted as the linewidth control capabilities of the particular exposure/resist technology (*5,17*). These data are subsequently plotted as a function of time and used to monitor the performance of a lithographic line, as shown in Figure 6. Cumulative data of this type is very important in maintaining optimum performance of a process line.

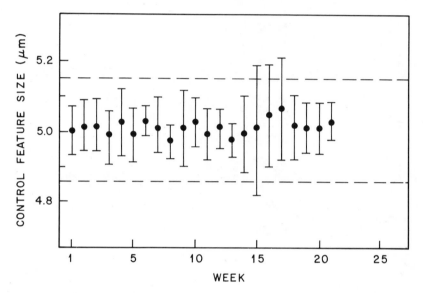

Figure 6. Linewidth control data for a typical process line. Weekly averages are shown for a 5.0 μm control feature. The dashed lines represent limits. Note weeks 15-17 represent a processing problem (etching in this case) that was corrected in week 18.

4.2.d Defect Density. From the standpoint of process development it is important to distinguish between the various types of defects [several of which are illustrated in Figure 7 (*18*)] that are normally encountered in semiconductor fabrication. Defects are imperfections in a pattern that result in an inoperative device. Such defects are called fatal defects as opposed to defects that are nonfatal. The latter are only cosmetic in nature and it is important that we be able to distinguish between the two. The

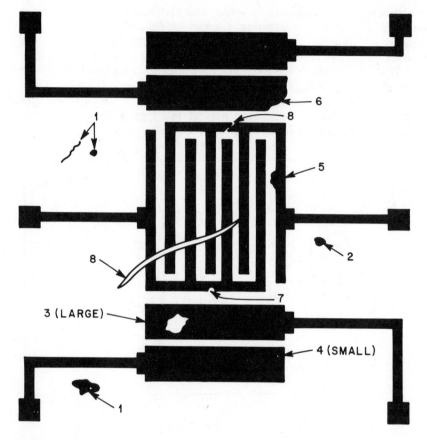

Figure 7 Types of defects commonly encountered on a process line. (1) contamination, (2) opaque spot, (3) large hole, (4) pin hole, (5) excess material, (6) lack of adhesion, (7) intrusion (mouse nip) and (8) scratch.

size, shape, and distribution of defects across a mask or wafer are frequently important factors in determining the *source* of defects in a process. It should be noted that a defect may reflect the *presence* or *absence* of material in patterned areas. For mask-related technologies, defects in a master mask pattern large enough to be resolved by the lithographic process, will be transferred to the wafer and cause defects in devices. Further, each lithographic step generates additional defects which all combine to reduce the total process yield of good devices. Price (*19*) proposed the following relation between device yield and defect density:

$$Y = \prod_{i=1}^{L} (1 + x_I)^{-1} \tag{5}$$

where Y is the potential yield of good devices or chips from a given mask set, $x_i = D_i A$ where A is the area of a chip and D_I is the average density of fatal defects on the ith mask level. Another expression relating defect density and device yield has been proposed by Murphy (*20*) and can be expressed in the form,

$$Y = \prod_{I=1}^{L} \frac{(1 - e^{-x_I})^2}{x_I^2} \tag{6}$$

to predict a cumulative yield for all L masks levels. If we assume $x_I = L$, Equation 5 can be rewritten to obtain

$$Y = (1 + x)^{-L} \tag{7}$$

and Equation 6 becomes

$$Y = (1 - e^{-x})^{2L}/x^{2L} \tag{8}$$

Figure 8 is a plot showing the effect of the defect parameter x on yield for 1, 5 and 10 mask levels. Note how rapidly chip yield decreases as the number of mask levels (L) increases.

Device defects stem from hardware, processes, and materials. Hardware-related defects include mask defects and contamination in the exposure environment. Those cauused by processing are determined by the cleanliness of the process and the number of processing steps. Material-related defects are caused by particulate matter in the resist or to the formation of unwanted insoluble particulate matter after the resist is coated and patterned. In summary, defects are generally caused by dirt and/or polymer particles, and great care must be exercised in eliminating unwanted contamination at every step in the lithographic process.

Quantitative inspection for defects has received much attention for both masks and wafers (*17,29*). Masks are much easier to inspect than wafers since one can use transmitted rather than reflected light. Human eyesight can resolve features 0.1 mm in size over a field of about 250 mm in diameter and hence, in order to inspect an entire mask for 2.0-μm defects, one must use 50X magnification and 35-40 separate glances over the field. A skilled inspector can inspect a 100 mm x 100 mm field in about two hours. This is an extremely time consuming process, and can involve errors resulting from fatigue of the human operator, which may result in there being as much as a factor of 4 difference in the total number of defects recorded on identical samples between two or more inspectors. Over the past several years many automatic defect inspection systems have become available. These systems can detect defects as small as 1.0 μm and require

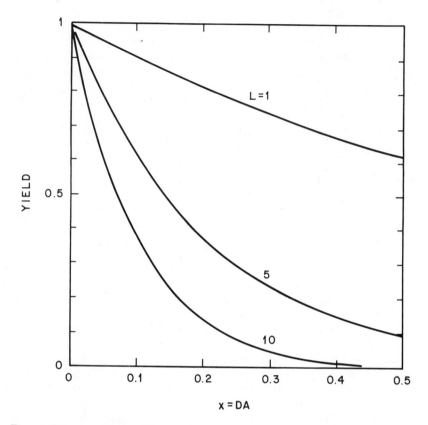

Figure 8 Device chip yield as a function of mask defects (using the defect parameter x) for 1, 5 and 10 mask levels. Most devices use 7-10 levels.

less than one-half hour to inspect a 4" mask. Since these units are under the control of a computer, it is easy to plot the distribution and size of defects over an entire sample. This type of reliable, quantitative information is very important in engineering a new process as well as monitoring an existing process.

4.2.e Etch Resistance. Etch resistance is the ability of a polymer to withstand an etching environment during the pattern transfer process and is perhaps the most difficult requirement to achieve for a polymeric resist. It also includes resist adhesion, which is an important requirement for liquid etching which is generally used to delineate features 2 μm and larger.

Most dry etching techniques rely on plasma-induced gaseous reactions in which the substrate is placed in an environment with a high radiation flux, frequently reaching temperatures in excess of 80°C. Whereas dry etching places less of a demand on adhesion, it does require a polymer to exhibit thermal and radiation stability. On the other hand, the requirement of high sensitivity dictates the use of a polymer that is unstable to radiation. This is especially true of chain-degrading positive resist systems whose design presents an obvious dichotomy to the resist chemist. Clearly then the physical and chemical properties of the resist represent fundamental limitations on etch resistance. The reader is referred to Chapter 6 for a detailed discussion of dry etching techniques.

4.3 Processing Steps

The quality and fidelity of the developed three-dimensional relief image in a resist is determined, to a large degree, by the physical and chemical processes which are involved in the various steps of the process development sequence outlined in Figure 1. In this section we will outline the process variables for each of these discreet steps and, where appropriate, describe the interactive nature with other steps. Table V lists the processing variables associated with each step that affect lithographic performance. Many of the specific examples given in this chapter regarding processing variables

TABLE V. Chemical and Physical Processing Variables
that Affect Lithographic Performance.

STEP	VARIABLES	
	CHEMICAL	PHYSICAL
CLEANING	Type of reagent Surface composition	Cleaning time Temperature Equipment condition
SPIN-COATING	Polymer composition Molecular weight Solvent boiling point Solvent vapor pressure Solution viscosity Solution concentration	Spinning speed Acceleration Temperature Humidity Filtration

TABLE V - Cont.

STEP	VARIABLES	
	CHEMICAL	PHYSICAL
PREBAKING	Polymer T_g and T_m Polymer stability Solvent composition	Temperature Time Atmosphere
EXPOSRE	Ambient gas in exposure chamber. Time after exposure	Energy density Exposure time Time after exposure Mechanical stability
POST-EXPOSING	Gas composition	Time Temperature
DEVELOPING	Developer composition Rinse composition Humidity Developer age (for dip) Batch size (for dip)	Spray: pressure and flow rate Temperature Time of developer, rinse and overlaps
PLASMA DESCUMMING	Gas composition	Reactor configuration Time Temperature Power Frequency Gas flow rates
ETCHING	See Chapter 6	See Chapter 6
STRIPPING (wet)	Reagent Age of bath	Temperature Time Degree of agitation Batch size

will refer to electron beam resists; however, the conclusions are completely general and equally applicable to any resist system. In Figure 1 the steps enclosed in dotted lines indicate optional steps or steps not required for all resists.

4.3.a Substrate Cleaning.   Many processing problems are frequently traced to dirty or contaminated substrates. Hence it is imperative that a substrate be atomically clean and free from any contamination in order not to adversely affect the lithographic process (*24*). The most common problems associated with dirty substrates are poor adhesion and defects such as pinholes and opaque spots. Poor adhesion leads to loss of linewidth control across a substrate (due to undercutting in liquid etching and or developing) and in the worse cases, to complete loss of pattern elements. The important variables associated with the cleaning step include type of reagent, the time and temperature of cleaning, type of surface to be cleaned, and choice of cleaning equipment.

4.3.a.1 Fundamental Considerations.   There are three types of surfaces normally encountered in semiconductor fabrication. Surfaces may be metallic, such as aluminum or gold semiconducting, such as silicon or III/V compounds, or insulating such as silicon dioxide or silicon nitride (*21-23*). A knowledge of the chemical composition of the surface is important, since it influences the type of contamination that can be attracted and suggests the particular analytical technique that can be used to determine the type of contamination. With the exception of semiconducting surfaces, all surfaces are formed by vacuum deposition techniques or thermal oxidation. It is important to note that in most cases the surface is in its most pristine or clean form immediately after deposition, and further cleaning can generally be avoided by coating the surface with resist immediately after deposition.

Contamination found frequently on surfaces can be conveniently considered in three categories: organic films, inorganic films, and particulate contamination (see Table VI (*24,25*)). Organic films may be subdivided into high molecular weight polymeric films and low molecular weight films comprised of oils and greases. Possible sources of low molecular weight organic contamination include contaminated deposition equipment, airborne organic materials from lubricants, air-handling equipment, or other machinery contained within the processing environment, as well as sweat, oil from personnel handling the substrates. Contamination in the form of polymeric films may stem from previous processing steps or contaminated wetting and adhesion promoters. Whenever plastic containers are used for the storage of substrates, it is possible for plasticizers to vaporize slowly from the container and deposit on the substrate. This is particularly true if the container and substrate have been exposed to elevated temperatures. Organic film contamination usually results in loss of adhesion of the polymeric resist film to the substrate and shows up as either severe undercutting in the etching step, or the loss of dimensional integrity in the resist pattern immediately after development.

Inorganic film contaminants are usually oxides, inorganic salts and/or water and are present as a result of reaction with ambient air, exces-

TABLE VI. Examples and Sources of Contaminant
Commonly Encountered in Microcircuit Processing

Contaminant Type	Specific Type	Possible Sources
ORGANIC FILMS	Oils or grease (low molecular weight)	Vacuum pump fluids Lubricants from air handling unit or other machinery Sweat and oil from personel
	Polymeric films (high molecular weight)	Plasma deposition from improper operation of sputtering equipment Deposits from previous resist processing steps Contaminated wetting agents or adhesion promoters. Plasticizers from storage containers.
INORGANIC FILMS	Oxide films	Air oxidation Excessive heating O_2 in deposition equipment Oxidative cleaning processes
	Salts	Airborne chemicals Storage containers Contaminated cleaning reagents
	Water	Humidity
PARTICULATE	Organic	Residual resist Contaminated cleaning reagents or adhesion promoters. Dirty spin coating equipment or drying ovens. Human debris such as dandruff, hair, etc.
	Inorganic	Airborne "dirt" Contaminated deposition or baking equipment Contaminated storage containers

sive heating of the substrate, residual oxygen in plasma deposition equipment, contaminated storage containers or baking ovens, and contaminated cleaning reagents. Frequently surface-adsorbed water is encountered.

Particles represent the most common form of contamination and are classified as either organic or inorganic. Organic particulate matter arises from several sources including residual resist from previous processing, contaminated reagents used for cleaning and adhesion promoters, dirty spin-coating equipment or drying ovens, and human debris such as dandruff and hair. Inorganic particles represent perhaps the most commonly encountered form of contamination and result from airborne dirt or contamination in deposition or baking equipment. This type of particle is usually large, irregularly shaped and is easily identified by low magnification optical inspection.

Various optical inspection techniques such as bright field, dark field or interference contrast optical microscopy may be used to examine the substrate for defect-causing contaminants. These techniques provide visual information such as particle shape, film thickness, etc. Scanning electron microscopy, low energy electron diffraction, and photoemission spectroscopy are also useful techniques for detecting all three types of contamination. If organic film contamination is suspected, techniques such as Fourier transform infrared analysis or Auger spectroscopy are useful in identifying the specific chemical composition of the organic material in question. Inorganic contaminants can be identified by x-ray analysis, atomic absorption spectroscopy or spark emission spectroscopy. It is beyond the scope of this work to go into detail on each of these analytical techniques, and it is recommended that competent, analytical scientists be consulted whenever chemical contamination is encountered as a problem.

4.3.a.2 Engineering Considerations. Once it has been determined that there is a contamination problem on the process line, one must then take steps to either *eliminate* the source of contamination or *clean* the affected substrates (*26,27*). It is certainly preferable to eliminate the source of contamination although this is not always practical. Certain types of contamination, for example, are inherent to previous processing. As a general rule though, eliminating contamination is a straight-forward engineering problem most frequently involving the cleaning of the process environment.

Under circumstances where substrate cleaning is necessary, four techniques are generally employed. These are:

1. Solvent treatment

2. Thermal treatment

3. Plasma or glow discharge techniques

4. Polishing

TABLE VII. Cleaning Procedures and Reagents

Technique	Reagents	Type of Contaminant Removed
SOLVENT	HF, HCl, HNO$_3$	inorganic
	H$_2$SO$_4$/dichromate Freons Ketones	organic
	Alcohols	H$_2$O
THERMAL	O$_2$ Vacuum N$_2$	organic SiO$_2$, H$_2$O H$_2$O
PLASMA (glow discharge)	O$_2$/H$_2$O Freons w/wo O$_2$	organic inorganic
POLISHING	Abrasive compounds	all

Table VII lists these four techniques along with appropriate reagents used with each method.

In solvent treatment reagents used include acids and bases as well as organic solvents such as freons, alcohols, ketones, etc. The solvent is used either to dissolve the unwanted contaminant or to convert the contaminant to another chemical species that can be removed easily in a subsequent step. For example, a typical contaminant encountered in silicon devices is unwanted SiO$_2$ films, which are formed on crystalline silicon surfaces as a result of air oxidation of the substrate. Hydrofluoric acid and certain plasma etching techniques are convenient methods for removing unwanted SiO$_2$. Other inorganic debris can be removed by strong acids such as hydrochloric or nitric acid followed by rinsing in copious amounts of deionized water. Glass cleaning solutions based on sulfuric acid and potassium dichromate as well as freons and ketones, are effective reagents for removing organic contamination. Needless to say, it is important that the chosen reagent not attack or modify the substrate in an undesirable manner. Alcohols are generally used to remove water or compounds that are soluble in water. It should be noted that the use of organic materials for cleaning substrates raises the possibility of depositing residues that are contained as impurities in the solvent. This is particularly true of ketones and freons which should be freshly distilled and filtered prior to use.

Many methods may be employed to apply the solvent to the substrate. These include: 1) vapor degreasing, 2) ultrasonic immersion, 3) pulsating spraying and 4) dipping. Vapor degreasing is a widely used and efficient method for cleaning semiconductor substrates. The substrate is suspended in a specially designed chamber so that solvent vapors condense on the surface and flush off adhering contaminants. With this method, the substrate is repeatedly washed in fresh, distilled solvent while the contaminants remain in the liquid phase at the bottom of the vapor phase degreaser. Liquid sprays are often employed in conjunction with vapor-phase degreasing to assist the removal of debris by mechanical action. Solvents used in vapor-phase degreasers include freons and chlorinated hydrocarbons. Flammable liquids should be avoided for safety considerations.

Ultrasonic cleaning represents one of the best and most efficient methods of cleaning *particulate* matter from substrate surfaces. This method involves submerging the substrate in the appropriate reagent and pulsating with acoustic vibrations of 20,000 — 50,000 Hz These high audio frequencies produce cavitation in the liquid in which thousands of microscopic bubbles per second are rapidly formed and collapsed. Cavitation results in a microscopic scrubbing action that removes not only adhering films but also particulate contamination. Ultrasonic cleaning is a rapid process requiring only a few minutes and lends itself well to automated production lines. Although it is an effective technique, there are several disadvantages which should be enumerated. First, the cleaning bath becomes contaminated since there is no mechanism for removing the contaminants after they are washed from the substrates. The bath therefore requires frequent filtration (or purification) to prevent "clean substrates" from being contaminated. Secondly, mechanical failure of the substrate film can occur as a result of the ultrasonic energy imparted during the cleaning cycle. This frequently results in film loss in certain regions and in an extreme case, the entire film may be removed from the substrate.

Pressure sprays pulsing at frequencies of 10 — 20,000 Hz offer an advantage over ultrasonic cleaning in that they avoid cavitation erosion and film failure, and yet are effective in removing even strongly adhering particles. Spraying also has the advantage of always using fresh reagents, although certain commercial units recycle the cleaning reagent several times.

Immersion cleaning or dipping is often employed, particularly in situations where the dissolution of the contaminant requires and inordinate period of time such as several minutes. In all cases where soaking is involved, it should be followed by some type of spray cleaner with fresh reagent, followed by thorough drying.

High temperature thermal baking is an effective technique for removing organic contamination, water, and occasionally very small amounts of silicon dioxide. Providing the substrate can withstand thermal cycles, baking at 1000°C in oxygen, vacuum, or a reducing ambient will result in almost perfectly clean substrates. After the diffusion steps are completed, however, this technique is usually avoided since device performance can be changed at such high temperatures.

Plasma cleaning using either oxygen, freons, or wet air represents a fast growing and effective method of cleaning semiconductor films. Many experimental and production equipment designs have been described in detail which are all similar to those used for plasma etching which is discussed in Chapter 5 of this work (see sections 5.2 and 5.6 for details of equipment and chemistry used in plasma cleaning). Plasmas have the advantage of offering a controlled and reproducible environment containing either reducing or oxidizing reagents in addition to chemical species which can convert a contaminant to a volatile material. Since plasma cleaning is accomplished in vacuum, it is inherently clean and usually does not introduce additional problems.

Polishing uses abrasive compounds and mechanical action to remove contaminants. Although used extensively in preparing glass substrates for chromium masks and in obtaining the desired flatness of silicon wafers, it is seldom, if ever, used once the silicon wafer enters a process sequence.

As we have seen there are a variety of techniques and reagents for cleaning a substrate which are available to the process engineer. It should be readily apparent though that one should strive to eliminate the source of contamination, rather than cleaning the contaminant itself after it contacts the wafer. While cleaning steps can remove an immediate problem, they represent additional processing steps which can possibly lead to even more contamination. Consequently, they should be used sparingly and their needs constantly reviewed. It is not unusual for cleaning steps to be introduced into a processing line and remain a part of the fabrication sequence long after their usefulness has expired, causing far more trouble than they are worth.

4.3.b Resist Coating. Coating the substrate with a radiation sensitive polymer is the first process step in the semiconductor manufacturing sequence involving the resist. Resist materials are usually high-molecular weight polymers dissolved in organic solvents and are deposited by a variety of techniques that are standard in the coatings industry. The objective of this step is to obtain a uniform, adherent, defect-free polymeric film over the entire substrate. Spin coating has long been accepted as the best coating method for obtaining the above properties. Spin coating is accomplished by flooding the substrate with a resist solution and rapidly rotating it at a constant speed between 1,000 and 10,000 revolutions per minute until the film is dry (*22-24,30,31*). The film uniformity across a single substrate and from substrate to substrate must be at least ± 15.0 nm in order to ensure reproducible linewidths and development times. Variables that affect the spin coating process are given in Table V and are divided into two categories: chemical and physical.

4.3.b.1 Fundamental Considerations of Spin Coating. The spin coating process consists of three steps:

1. Flooding the substrate with resist solution

2. Acceleration to a desired RPM and

3. Spinning at a constant RPM to near dryness

The initial resist application can be accomplished either by flooding the entire wafer with resist solution prior to spinning or depositing a small quantity of resist in the center of the wafer or substrate, spinning at low RPM (to produce a uniform liquid film), and then accelerating to the desired final spinning speed. Both approaches are designed first to obtain a liquid film of uniform thickness and subsequently to dry it at a constant rate to yield a uniform, solid polymeric film. During stage 3, the resist solution experiences two forces — an outward centrifugal force caused by rotation and a shear force on the flowing liquid annulus (Figure 9). Once the uniform liquid film has developed, the remainder of the process consists of solvent evaporation to produce the solid film. Fundamental parameters which govern film uniformity and thickness include polymer composition, molecular weight, solution concentration (viscosity) and, angular velocity and acceleration of the spinner (*32,33*). Additionally, the physical and chemical properties of the solvent are important, as we shall see later.

For a given resist system with constant molecular weight, solution concentration, etc, the film thickness depends only on the spinning parameters. Figure 10 is a spinning curve which shows the resist film thickness (t) as a function of rotating speed ω for three resist solutions. These samples cover a range of molecular weight and concentration and were formulated to have a constant solution viscosity (*28*). The log t vs. ω plot is linear in all three cases with a slope independent of the concentration or molecular weight and suggests the following relationship where both k and α are constants

$$t = k\omega^{\alpha} \tag{9}$$

The constant α is determined from the slope of the line.

Figure 11 shows the spinning curve for a resist as a function of spinning speed for three concentrations (this is the same polymer as used in Figure 11). Note that the slope of these curves is the same as in Figure 10. It

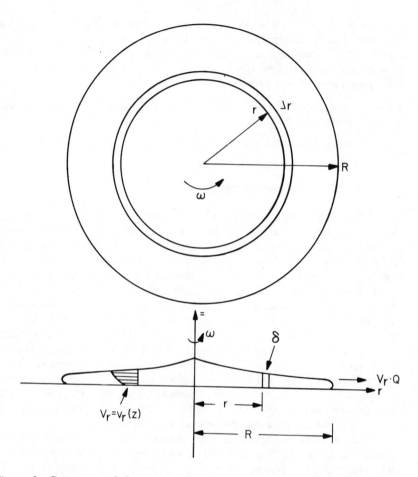

Figure 9. Symmetrical flow pattern of a homogeneous liquid on a rotating disk showing flow rate, Q and velocity profile $v_r(z)$. (Reproduced with permission from Ref. 27).

is evident from these plots that at constant spinning speed, the film thickness increases with concentration. This is shown quantitatively in Figure 12 which plots log t against log C where C is the polymer concentration in grams per 100 ml solution. The curve is linear, suggesting the following relationship:

$$t = k'C^\beta \qquad (10)$$

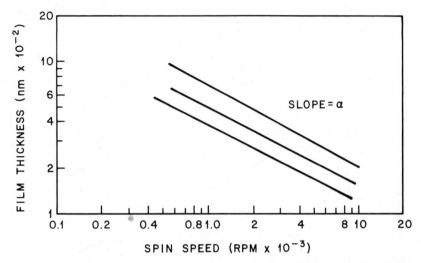

Figure 10 Resist film thickness as a function of spinning speed (in RPM) for three different molecular weights of the same polymer. The solution concentration is constant for all samples.

where both β and k' are constants. Again, β is determined from the slope of the line in Figure 12.

Molecular weight (measured by intrinsic viscosity) has a dramatic effect on the film thickness at constant rotational speed. Figure 13 shows the spinning curve for several polymer solutions at a constant solids concentration and varying intrinsic viscosities (i.e., molecular weight). Note the slope is again the same as for Figures 10 and 11. If one constructs a log-log plot of film thickness as a function of intrinsic viscosity at a constant spin speed, a linear curve is again obtained as shown in Figure 14. We may therefore write

$$t = k''[\eta]^\gamma \tag{11}$$

where k'' and γ are constants. Equations 9, 10 and 11 can be combined to give the following empirical relationship that describes the effect of molecular weight, solution concentration, and spinning speed on the resist film thickness t,

$$t = \frac{KC^\beta[\eta]^\gamma}{\omega^\alpha} \tag{12}$$

where K is an overall constant. Once the various constants have been determined for a given polymer and solvent system, the equation can be used to predict the film thickness for various molecular weights and solution concentrations.

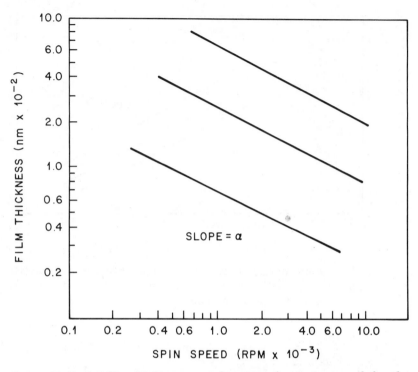

Figure 11. Resist film thickness as a function of spinning speed for three solutions of different concentrations. The polymer composition and molecular weight are constant for the three samples.

      While Equation 12 allows us to predict the thickness of the film which can be spun from a given resist solution, it does not predict the *quality* of the film that will be obtained, nor does it address the variation in resist film thickness across a wafer or from wafer to wafer. These properties are determined primarily by the spinning solvent. The choice of spinning solvent is largely an empirical exercise and is determined by considerations of polymer-solvent interaction and solvent volatility. Polymers exist in solution, not in an extended chain conformation, but as random coils whose size is characterized by the mean square end-to-end distance known as the radius of gyration. Solvents are characterized as thermodynamically "good" or "poor" depending on the size of the polymer coil in solution. In a good solvent, the coil is relatively expanded corresponding to a large value of the radius of gyration with a corresponding large value of polymer-solvent

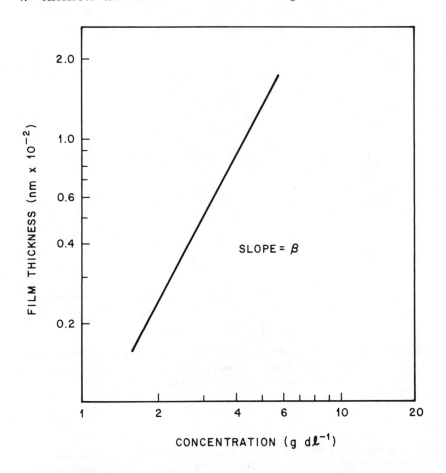

Figure 12. Resist film thickness as a function of solution concentration at constant spinning speed (1000 RPM). If other spinning speeds are plotted (from Figure 11) a family of curves will be generated all of which exhibit a slope equal to β.

interaction parameter. As the solvent becomes progressively poorer, the coil shrinks until finally the polymer and solvent are no longer miscible and the polymer precipitates. Generally a solvent becomes poorer with decreasing temperature until finally a point called the θ-temperature is reached below which precipitation of the polymer will occur. The large chains usually precipitate first with chains of lower molecular weight being precipitated on further cooling. This phenomenon forms the basis of a technique for poly-

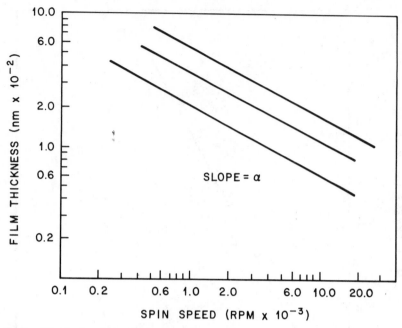

Figure 13. Resist film thickness as a function of spinning speed for three samples of different molecular weight. The solution concentration and polymer composition are the same for all samples.

mer fractionation. We should note that some solvent systems display a lower critical solution temperature, i.e., the θ-temperature is approached as the temperature is increased.

It is clear from this discussion that a solvent should be chosen which has the correct solution properties to yield a uniform film. The polymer should remain in solution throughout the entire spinning phase; otherwise aggregation may occur which can lead to opaque spots, pinholes or irregular lines after development.

Solvent volatility (vapor pressure) is another parameter which has a marked effect on film uniformity. Solvents which are highly volatile evaporate rapidly resulting in too rapid an increase in viscosity thereby preventing the film from forming with uniform thickness over the whole of the substrate. The problem becomes more severe as substrate size increases. Solvents with too low a volatility may result in drying times which are too long and increase the probability of particulate matter sticking to the film and creating a potential defect site.

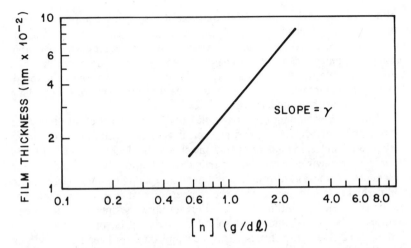

Figure 14. Resist film thickness as a function of polymer molecular weight ([η], intrensic viscosity) at constant spinning speed (1000 RPM). Other spinning speeds may be used from Figure 13 and a family of lines will be generated all with a slope equal to γ.

These considerations form guidelines for solvent choice which can be aided by consulting handbooks of chemical information. In many cases, combinations of different solvents may be needed to achieve the desired result.

4.3.b.2 Engineering Considerations. In addition to the fundamental considerations discussed above, certain engineering considerations must be understood in order to obtain defect-free resist films of the required uniformity. Once a resist solution is prepared, it is necessary to obtain a spin curve which describes the effect of RPM or spinning speed on thickness as shown in Figure 10. This curve is generated by spinning a series of substrates at spinning speeds over the range of 1,000 - 10,000 RPM, prebaking these substrates, and subsequently measuring the thickness of each film. Thickness measurements may be made by several techniques, such as interference microscopy, automatic film measuring devices, or stilus instruments (*29*).

Defects may be generated during this spin coating operation in which case they are generally observed to be distributed radially. Random defects may also be generated but these come from other sources and not the spin coating operation itself. The engineering considerations which are related to hardware and which govern film uniformity and defect density include the cleanliness of the environment (including resist dispensing equipment), temperature control, humidity control, mechanical integrity of the spinner (including constant acceleration and speed), exhaust equipment. Reliable spinning equipment can be purchased from several manufacturers and should be capable of being calibrated for both acceleration and spin speed. Further, the equipment should maintain calibration over relatively long periods of time without undue drift.

During the spinning operation, the polymer film is sticky or tacky for a considerable period of time, and as such is very susceptible to entrapment of airborne particles. Spinning operations should be done in at least a Class 100 or better environment, with particular attention being paid to the maintenance of the air filtration equipment and spinning hardware. Since the film thickness is dependent on the temperature, it is important that the latter be controlled to within $\pm 1°C$. Furthermore the spinning equipment, substrates to be coated and resist solution should all be in thermal equilibrium with each other. This is easily accomplished by keeping everything within the same environment for at least 24 hours prior to spin-coating. During the spinning operation, solvent evaporation results in cooling of the substrate, leading to possible condensation of water vapor. To avoid this problem, the humidity should be controlled at less than $30\% \pm 2\%$ RH. All resist systems use organic solvents, and particular care should be paid to exhausting the solvent fumes that result from the evaporation part of the spinning cycle. The exhaust equipment and associated electronic hardware should be spark-resistant to avoid the possibility of fire.

One of the major sources of problems encountered in spin-coating hardware is the dispensing equipment. Frequently it has been recommended that the resist be dispensed directly from the bottle thereby avoiding all mechanical pumps, syringes, or additional filtering equipment. Provided the resist solution has been adequately filtered prior to dispensing, direct application from the bottle should yield the lowest defect films possible.

There are several engineering considerations regarding the resist material. It is mandatory that the resist composition (both of the polymer as well as the spinning solution) be maintained within tight specifications and that the resist be free of all particulate matter above 0.2 μm in diameter. This may be achieved by filtration through standard porous Teflon filters for which several commercial systems are available. One should be careful to avoid additional chemical or particulate contamination of the solution, or degradation of the polymer during the filtering operation. It is also important that the resist solution be free of all entrapped air as an air bubble can cause a defect which is similar in appearance to that caused by a particle. One should generally allow the resist to sit several days after filtration and prior to application, to allow all dissolved air or other gases to escape.

Once a satisfactory film has been obtained, the coated substrate should be transferred to a suitable storage container and moved directly to the subsequent processing step, viz., prebaking. It is important that unbaked, freshly spun films not be stored for periods of time in excess of a few hours at this stage, since many resists are particularly vulnerable to particulate contamination. Once particles contact a resist surface prior to prebaking, they are almost impossible to remove and will cause opaque spots or pinholes after exposure and development.

4.3.c Prebaking.   Before we consider prebaking, it is important to understand the physical and chemical state of the polymer film after the spin-coating process. After spin-coating and air-drying, the polymer film contains 1 to 3% residual solvent and may contain built-in stresses caused by the shear forces encountered during the spinning process. In order to remove residual solvent and anneal any stress in the film, one must use a baking step prior to exposure. Residual solvents will adversely affect both subsequent exposure and developing cycles, and stress in the film may result in loss of adhesion or erratic developing or etching during subsequent processing. Prebaking is frequently considered a trivial step; however, it is very important that it be done properly, since it ensures that the resist film has uniform properties both across the wafer or substrate and from one substrate to another. It therefore ensures reproducible processing of each wafer and wafer lot. If the prebaking schedule is followed rigorously, this step does not generally introduce any problems in subsequent processing unless of course the baking equipment itself is contaminated in which case, defects may be generated. It is important to note that problems caused by an incorrect prebaking cycle are not fully realized until after the developing step. Consequently, in order to specify prebaking schedules, it is necessary to have the developing cycle and exposure conditions optimized to some extent. Relevant variables that affect prebaking conditions are given in Table V, and will be discussed in the next two sections.

4.3.c.1 Fundamental Considerations.   Polymeric resists are generally amorphous polymers with their physical properties being determined by the degree of molecular motion allowed within individual chains. At elevated temperatures where viscous flow is possible, there is considerable motion of chain segments. As the temperature is reduced, the motion of these segments becomes increasingly hindered and is eventually halted at a particular temperature at which the polymer has the characteristics of a glass. If this temperature, known as the glass transition temperature (T_g), is at or below room temperature, the polymer is considered to be a rubber while if it lies above room temperature, the polymer is considered to be a glass (*31,36*). Both states are encountered in resists. Upon cooling from the melt, some polymers form crystals at a characteristic temperature known as the melting point, T_m. Such polymers are not generally useful as resists since the formation of crystalline segments prevents the formation of uniform isotropic films. Above T_g, the polymer chains become mobile, at which point the polymer begins to flow under small amounts of stress. Thus heating the resist film above its glass transition temperature for a reasonable period of time enables the film to anneal into its most stable energetic state. Further, since diffusion is an activated process, solvent removal is also facilitated. Sufficient time must be allowed for solvent molecules to diffuse out of the polymer matrix.

Resist materials are designed to be radiation-sensitive and as such are usually thermally sensitive. Consequently it is important that excessive baking temperatures that would degrade or induce undesired chemical reactions in the polymer not be used. This implies control of the time and temperature which are the main prebake variables. Although optimization of these variables is done somewhat empirically, there are certain analytical techniques useful in guiding one to initial prebaking conditions. The required temperature and time can be determined to a first approximation, using thermal methods of analysis such as thermogravimetric analysis (TGA) or differential thermal analysis (DTA) (38). Thermogravimetric analysis is a technique that records the weight loss of a polymer as a function of temperature and indicates the temperature at which thermal degradation (i.e., volatization) occurs. A typical trace is shown in Figure 15. Differential thermal analysis may be used to determine the glass transition temperature and melting point of a polymer. A typical DTA trace of a polymer is shown in Figure 16. The thermal properties of the positive electron beam resist poly(butene-1-sulfone) (PBS) have been extensively investigated (39) using these techniques, and these data were used to establish thermal processing limits.

DTA is also useful in detecting chemical reactions in a polymer that occur without weight loss, e.g., crosslinking. Using these techniques, one can determine the minimum effective prebaking temperature and the upper temperature limit where incipient polymer degradation occurs. Thermogravimetric analysis can be used to follow solvent removal also and is useful in predicting the time required at a given prebake temperature to ensure complete solvent removal. Once a time and temperature have been established for prebaking, it is then necessary to evaluate the pattern quality and the etching characteristics to ensure that the prebake conditions were satisfactory. Several iterations of optimizing these variables is generally required before a satisfactory prebaking schedule is determined. It is also important to choose times and temperatures that allow some flexibility in prebaking conditions without adversely affecting the lithographic performance of the material.

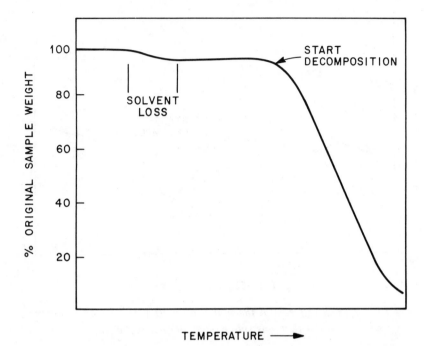

Figure 15. A typical thermogravimetric analysis (TGA) trace of a polymer resist that has been spin-coated and NOT prebaked. The small weight loss at low temperature is due to solvent evaporation and the high temperature weight loss is due to thermal decomposition.

4.3.c.2 Engineering Consideration. The primary engineering considerations in prebaking are the selection and maintenance of baking ovens. Several types of ovens such as forced air convection ovens, conduction ovens, and infrared heated ovens are commonly employed. The main requirements for baking hardware are:

1. Achieving a uniform temperature in a short period of time

2. Temperature control and uniformity

3. Cleanliness

As previously stated, it is important that the baking step not introduce undesirable particulate defects in or on the resist film. Since the resist is above its softening point or T_g during baking, any particulate matter con-

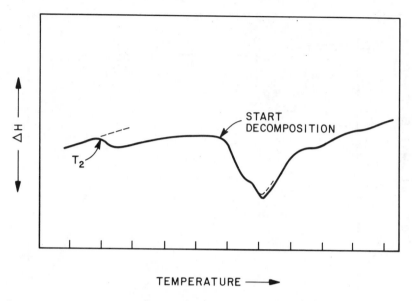

Figure 16. *A typical differential thermal analysis (DTA) trace of a polymer resist that was spin coated and not prebaked.*

tacting the surface will become embedded in the surface and will be nearly impossible to remove subsequently. Several vendors market convection ovens (forced air) that use internal filters to keep the oven ambient at Class 100 conditions.

It is important that the temperature of the oven be measured at the substrate location and not at some remote point in the oven. A frequent problem encountered in the prebaking step is incorrect temperature at the substrate. This is easily checked by attaching a thermocouple to a substrate of the same size and shape as those being processed, placing it in the oven and monitoring the temperature as a function of time. This should be done on a regular basis to ensure proper calibration of the baking hardware. Many ovens do not have the thermal capacity to heat large masses of material and bring them to thermal equilibrium over a short period of time. Under these circumstances the temperature will be dependent upon the number of substrates prebaked simultaneously. Thus for the particular oven in use, it is important to know if the temperature at the surface of the substrate is the same when several substrates are baked simultaneously.

4.3.d Exposure. After the substrate has been coated with resist and properly prebaked, it is next exposed to some form of patterned radiation to create a latent image in the resist layer. For the purpose of exposure discussions, it is important to differentiate between beam forming systems such as ions or electrons and mask systems as used in electron image projection, photolithography, and x-ray lithography. The quality of the latent image is dependent upon the exposing hardware, the physics and chemistry of the interaction between the radiation and the resist and, in the case of mask systems, the quality of the mask. In the following section we will discuss the processing variables affecting lithographic performance that are associated with the exposure step.

4.3.d.1 Fundamental Considerations. We recall from Chapters 2 and 3 that the chemistry and physics of exposure determine in large part the resolution of the latent image. In addition there are several processing considerations that must be taken into account. These include energy density and uniformity, dose rate, exposure time, and the exposure ambient (vacuum, oxygen, air, etc.). The energy density available from a source dictates the exposure time and resist sensitivity required to form the latent image. Clearly, a source with low energy output results in excessively long exposure times. It is also important that the energy source be stable both temporally and spatially during exposure. It has been shown in Section 4.2.a that the feature size is critically dependent upon the exposure dose and, hence, small variations in absorbed dose can create unacceptable variations in linewidth. Too high a dose rate can result in loss of reciprocity in the resist exposure. For example, it has been shown that reciprocity failure occurs at high dose rates during electron beam exposure of vapor developing resists such as poly(2-methyl-1-pentene sulfone) as seen in Figure 17. Further, marked differences in vapor development rate occur, depending on whether the dose is received in a single pass (assuming a fixed writing rate) of the beam or cumulated over several passes (Figure 18). The latter is considerably more efficient and is believed due to the fact that the unzip time for depropagation is finite relative to the dwell time of the beam.

The exposure ambient is also an important parameter. Electron beam and ion beam hardware must operate in hard vacuum. Hence reactions whose rates are affected by or perhaps dependent upon ambient gasses such as oxygen and water vapor will proceed at different rates in vacuum than in exposure environments where these gasses are present. Some photoresists and x-ray resists require an atmosphere containing known amounts of moisture or oxygen and it is important that the environment be reproducibly controlled both during and after exposure. Many solid state reactions in polymers continue to occur in vacuum for many hours after removal from the source or at least until quenched, usually by exposure to either oxygen or water vapor. This phenomenon is known as post-exposure curing and is

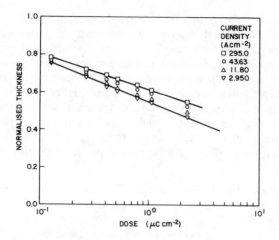

Figure 17. Effect of beam current (dose rate) on the vapor development of poly(2-methyl-1-pentene sulfone).

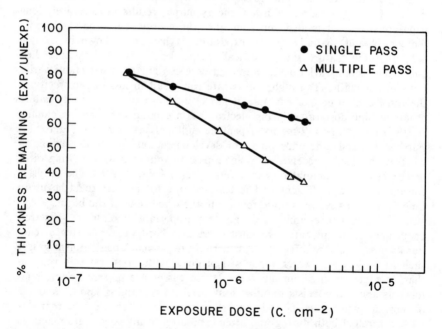

Figure 18. Effect of dose delivered in a single pass (at varying beam current) and multiple pass (at fixed current) on vapor development of poly(2-methyl-1-pentene-sulfone).

mainly seen in negative resists such as COP and PGMA (42-43) (see Section 4.3.b.1 for further discussion of this phenomenon).

Two component, positive photoresists (see Section 3.5.b) represent systems with unusual exposure characteristics caused by the standing wave effect (see Section 2.1.f) and "bleaching" or change in optical density during exposure (see Sections 3.5 and 3.9). Both of these phenomena result in nonlinear exposure throughout the thickness of the resist film, and result in uneven developing rates as a function of film thickness, making evaluation of these systems difficult.

4.3.d.2 Engineering Considerations. The charge deposited in the film and substrate during electron or ion beam exposure must be continuously removed so that the beam is not adversely deflected or affected by the charged substrate. One normally relies on the conductivity of the underlying substrate to achieve this, although with thick resist films it may be necessary to use a conducting layer or make use of the conductivity of the resist itself in order to achieve precise image placement. As mentioned in Chapter 2, it is important that the source be stable and uniform so as to deposit the same dose at every pixal element on the pattern. Any nonuniformity in dose will result in an error in the feature size of the latent image.

It is very important that the substrate be held mechanically rigid with respect to either the mask or the beam since any vibration or mechanical movement of the substrate relative to the source will result in misplacement of the image. It is also obvious that the longer the exposing time required, the greater are the demands placed on mechanical stability. Great strides have been made in the last ten years in vibration isolation and mechanical stability, and generally speaking these factors do not limit the lithographic performance of the hardware. Laser interferometers are extremely useful in monitoring the precise position of the substrate relative to the mask or beam.

As in all processing steps, cleanliness of the exposure hardware is of paramount importance. Any particle that lands on the resist prior to exposure, will shield the film underneath the particle from the exposing radiation and give rise to opaque spots in the case of positive resist, or pinholes in the case of negative resists. Particulate contamination is especially troublesome with electron beam and ion beam systems where the probability of a particle landing on a substrate is increased relative to other techniques because of the much longer exposure times involved.

4.3.e Post-Exposure Treatment. After the polymeric resist film has been exposed and the latent image generated, there may be chemical reactions that continue to occur after exposure. As a consequence, it is often necessary or advantageous to control the environment of the exposed polymer film in order to allow the desired reactions to continue to completion or induce new reactions that aid in developing subsequent relief images in the resist film. Several post-exposure treatments have been reported in the literature and are summarized below:

1. Thermal

2. Flood exposure with other types of radiation

3. Treatment with reactive gas and

4. Vacuum treatment

Methods 1 and 3 have been utilized in dry developed resist systems. To our knowledge, there are no resist systems commercially available that depend on post-exposure treatment other than the post-curing effect in negative electron beam resists mentioned earlier. Since such systems are still largely in the research phase we will not discuss them here but rather refer the reader to the literature for more detailed descriptions (44-50).

4.3.e.1 Fundamental Considerations. When high energy radiation is absorbed in a polymeric film, reactive species such as radicals, radical ions or ions are created which initiate chemical reactions that either alter the molecular weight of the polymer or change the chemical composition of sensitizer molecules. These active sites may be quite stable, particularly if the film is below T_g. Not only that, but the rates of reaction in solid state polymers are frequently slow, and although the initiation reactions may be quite fast, the activated molecular species may continue to diffuse through the polymer matrix long after removal from the source causing additional reactions and further altering the solubility of the polymer. An example of this type of phenomon is the post-exposure curing that occurs with negative electron resists with low T_g's and/or a chain reaction mechanism for crosslinking. This is illustrated in Figure 19 where the crosslinking reaction is seen to continue for a considerable period of time.

This problem leads to two undesirable phenomena: first, since an electron beam exposure system requires 10 to 60 minutes in order to serially expose a wafer or mask, sections of the surface exposed last will show a degree of crosslinking *less* than sections exposed earlier. Since the linewidth of an image is proportional to the exposure dose, it is imperative that we allow the reaction to proceed to the same extent across the entire wafer if we are to control the linewidth across a wafer or mask.

4.3.e.2 Engineering Considerations. One of the primary reasons that resist systems employing complex post-exposure reactions have not gained wide acceptance commercially is that as a rule, these reactions are difficult to control and reproduce in a production environment. Although the reaction times or lifetimes of the active species are long compared to the exposure times, they are short when compared to overall process cycles. The reactive sites created in a solid polymer also continue to decay via side reactions, and if we are to reproduce the exact extent of reaction in a latent image, it is necessary that we be able to effect the precise degree of chemical reaction both across a substrate and from substrate to substrate.

Post-exposure curing is fairly straight forward and easily controlled. In the case of negative electron beam resists, it is conveniently controlled simply by leaving the sample in the vacuum environment until all areas on the substrate have had a chance equilibrate to the same approximate degree of reaction. Referring to Figure 19, it is clear that after approximately one hour exposure time in vacuum, the rate of reaction has decreased to a very low value. Therefore it is generally sufficient to allow the resist to reside in the vacuum chamber for a time approximately equal to two to three times the exposure time. By comparison, the other post-exposure procedures alluded to earlier are considerably more complex and will require extensive process development before they appear on the production line.

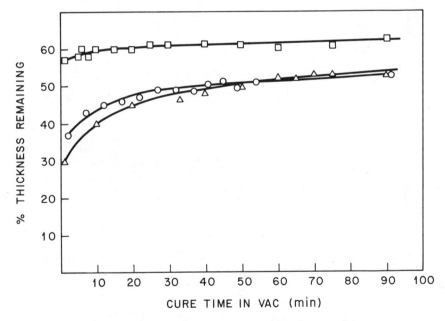

Figure 19. The fraction of film remaining after electron beam exposure and development as a function of vacuum curing time for three negative electron beam resists.

4.3.f Developing. Once the latent image has been formed in the polymer film, it must be developed to produce the final three dimensional relief image. The developing process is the most complex of the processing steps, has the greatest influence on pattern quality, and requires much more process development time than any other step. Two development processes have been used, viz., liquid development and dry development (plasma). Dry development of resists has been reviewed comprehensively by Taylor (*44-47*). It is still in the research and early development phases and will not be treated here.

There are two liquid development process which are employed: spray and immersion. Of these, spray development generally gives better pattern uniformity and process reproducibility, largely as a result of better time and temperature control and the continuous use of fresh developer. Table V lists many of the variables associated in the developing step all of which are interactive making optimization an extraordinarily tedious task. It is important to reiterate, however, that it is this step that has the greatest effect on the quality and appearance of the resist image; success or failure of the entire resist program may well depend on the successful accomplishment of this task.

In order to optimize the developing process, it is necessary to have a set of developing criteria such as the following that are typical of the criteria used in our own positive resist development work:

1. No reduction in original film thickness

2. Minimum developing time (less than one minute)

3. Minimum distortion and swelling during the developing process and

4. Faithful reproduction of the required dimensions

Development criteria for negative resists are essentially the same, although in this case, the original film must by necessity, be completely removed. The film thickness remaining is also never the original film thickness for negative materials since this condition would not result in the correct feature size, (see Section 4.2.a.).

4.3.f.1 Fundamental Considerations. During normal dissolution, solvent molecules penetrate into the polymer structure by a diffusion process producing a swollen surface layer between the solvent and virgin polymer. In the case of positive resists, minimum distortion implies minimum swelling in the regions immediately adjacent to the irradiated area which is removed by dissolution. For negative resists it implies minimum swelling of the irradiated region during extraction of the soluble (sol) fraction from the crosslinked gel.

Normally the suitability of a solvent is discussed in thermodynamic and kinetic terms where the kinetic influence is reflected in the value of the diffusion coefficient while the thermodynamic suitability influences the thickness of the swollen layer (*51-54*). Very little work has been done on quantifying the effect of these parameters on the kinematics of polymer dissolution. Tu and Ouano (*55*) described the dissolution of glassy polymers by a phenomenological model in which they showed that the rate of dissolution and the extent of swelling depend markedly on the concentration dependence of the diffusion coefficient of the solvent in the polymer. However, for most systems the relevant material parameters are not known, and the choice of developing solvent remains largely an empirical problem. Figures 20 and 21 show some examples of how solvent choice can influence significantly the quality and appearance of the resist. Temperature and humidity may also have a drastic effect on the solubility of a polymer in a given solvent or solvent mixture, and variations in these parameters can produce widely differing processing results, particularly when one is trying to differentiate between polymers with relatively small molecular weight differences. Positive resists, for example are usually given a dose necessary to effect minimal separation of the molecular weight distributions, and we must choose solvents that are only marginally poor for one region relative to the other. Figures 22 and 23 show patterns developed in some typical electron beam resists and illustrate the large variation in pattern quality as a result of rather subtle changes in the developer. One problem with very high resolution negative electron resists is shown in Figure 24, which shows resolution of the address structure in irradiated areas. The contrast of the resist is sufficiently high that the overlap region between adjacent scans receives a dose that is insufficient to cause significant gel formation. The "gaps" between the scan lines are not obvious during development, because neighboring lines are considerably swollen. However, if the sample is immediately quenched in the resist, the lines retract, giving rise to the so-called "swiss cheese" effect. (Complete separation is impeded by considerable adhesion between adjacent lines.) This effect can be eliminated by judicious choice of rinsing conditions aimed at de-swelling the polymer gradually (generally by employing a series of solvents of decreasing strength) or by post-baking schedules. Two component positive resists, based on novolac resins are developed using aqueous base (alkaline) solutions or liquid organic bases. Since these resists are very low molecular weight (see Section 3.5.b) they dissolve with a minimum of swelling (layer by layer) rather than by forming a thick gel layer first. This is the primary reason these systems exhibit high resolution. The required developer strength is determined by measuring the rate of dissolution of unexposed resist at various developer strengths. The developer strength is varied by diluting a strong, buffered base solution. The aqueous developers are usually potassium or sodium hydroxide solutions, buffered to the desired pH value.

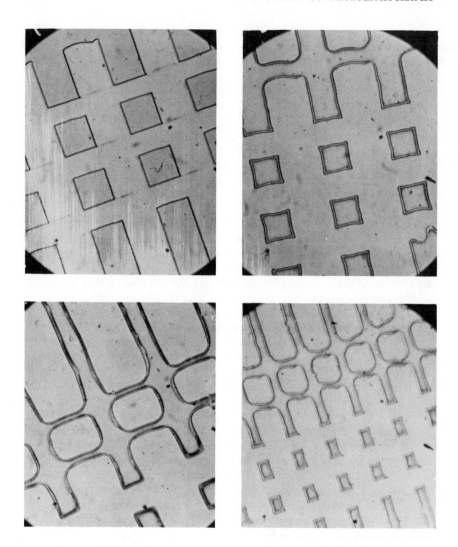

Figure 20. Optical micrographs of PBS resist patterns developed in four different ketone-based developing solvents.

In the final analysis, there is no substitute for a rigorous empirical evaluation of many solvents in order to find the optimum developer and rinses. In many cases mixtures of solvents prove to be beneficial.

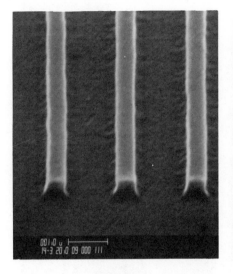

Figure 21. Scanning electron beam micrographs of COP negative electron resist patterns developed in two different ketone-alcohol mixtures.

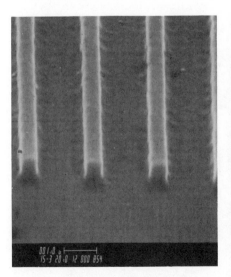
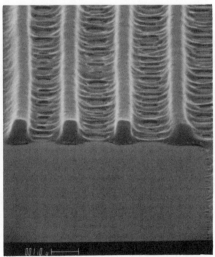

Figure 22. Scanning electron beam micrographs of a typical negative resist developed in different solvents. The dose was similar in both cases.

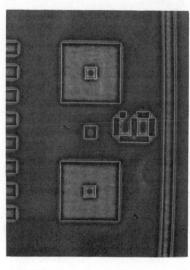

Figure 23. Effect of developing solvent on pattern quality of poly(butene-1-sulfone): (A) and (B) - 1 and 3 minutes in developer A; (C) and (D) - 1 and 3 minutes in developer B.

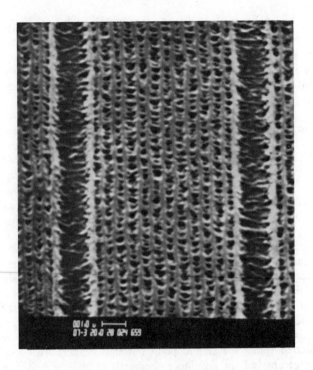

Figure 24. *"Swiss Cheese" effect in GMC negative resist produced by "shock" rinsing after development.*

4.3.f.2 Engineering Considerations. The primary requirement of the developing hardware is that it uniformly apply the developing solvents under reproducible conditions of time, temperature, concentration, and cleanliness. As with spin-coating, during the developing cycle the polymer is especially susceptible to particulate contamination, and it is difficult -- if not impossible -- to remove particles embedded into the swollen polymer matrix. Normally, the developing process is carried out in a Class 100 clean environment with temperature and humidity maintained within tight tolerances. Temperature should be maintained to ±0.5°C, and humidity ±2% RH. This is achieved by using temperature- and humidity-controlled cabinets that contain all of the developing hardware and chemicals.

It is also important that the developer be applied uniformly across the entire substrate for a precise length of time. This is quite easily achieved with dip developing, while in the case of spray development one must be very careful in adjusting and arranging nozzles in order to achieve development uniformity. One word of caution with spray developers is that during the spraying process, the temperature of the developer can vary because of evaporative cooling from the nozzle, and if one develops a series of wafers or substrates, the temperature may continually decrease to some equilibrium value over a period of some minutes. This means that the effective developing temperature of the first substrate will be higher than for substrates developed subsequently and will result in non-reproducible linewidths. This can be controlled easily by carefully monitoring the temperature and allowing the spraying system to come to thermal equilibrium before a substrate is introduced. Most commercial developing systems are controlled by microcomputers; thus reproducing the spraying times and overlap times of developers and rinses is relatively straightforward. It is necessary only to determine the optimum times for each operation. Immersion development, on the other hand, is more difficult to control, since this is normally a manual operation which is dependent on operators.

After the completion of the developing cycle, it is necessary to dry the substrate thoroughly of all residual developers and rinses. This is normally accomplished with a stream of dry air or nitrogen while spinning at several hundred rpm. We should point out that evaporative cooling causes both the substrate and the developer chemicals to become quite cool (sometimes as low as $10^{\circ}C$), rendering the substrate susceptible to water condensation. Generally, the last rinse used in the developing sequence consists of a chemical such as propyl alcohol that will remove water. It is also useful to employ heated, thoroughly dry nitrogen or air to blow off the substrate in order to ensure that no water condensation occurs during the drying process. If water condensation does occur, it can cause defects during the subsequent etching steps. Chromium masks are especially susceptible to water condensation, giving rise to opaque spots during etching.

After development, the pattern should be inspected thoroughly for both defects and linewidth control. Further, pattern quality on a process line should be continuously monitored after the developing step in order to ensure that the desired quality is consistently maintained.

4.3.g Plasma Descumming. The purpose of the developing process is to remove completely any unwanted resist in areas to be etched subsequently. Unfortunately, it is impossible in many instances to remove all resist during the developing step. This can occur with negative resists which tend to form a thin crosslinked film at the resist-substrate interface. It is frequently observed that after developing, the edges of some resists tend to be somewhat ragged and non-uniform as a result of swelling and deformation during

Figure 25. Effect of plasma descum on pattern quality.

developing or perhaps backscattering or proximity effects. Both of these problems can be overcome to some degree by using a mild plasma treatment after developing and prior to postbaking. It is the purpose of this plasma treatment to remove in a delicate, almost surgical-like manner, very tiny quantities of resist in unwanted areas. Resist removal also occurs in the patterned resist areas but as long as only a few hundred angstroms of total resist is removed, this is usually of little consequence. Figure 25 shows a typical negative resist pattern before and after plasma de-scumming. The improvement in pattern quality is obvious. Plasmas may also be used to trim or tailor the linewidth of resist features a few hundred angstroms at a time in order to achieve the correct dimension prior to etching.

4.3.h Postbaking. The last step prior to etching is postbaking which is designed to remove residual solvents and anneal the film in order to promote adhesion. The limitations and fundamental considerations for postbaking are the same as those discussed in Section 4.3.c under prebaking, with one exception. Since we have carried out the requisite radiation reactions in the polymer, it is less important that thermally-induced reactions be considered during the postbaking step, and it is frequently possible to bake at much higher temperatures and times to achieve the requisite lithographic performance. One important limitation is that the postbaking temperature must be less than that which causes the polymer to flow or melt. In addition, temperatures that will affect the underlying substrate adversely should be avoided. The postbaking cycle should be designed so as to have as large an operating window in both temperature and time as possible.

4.3.i Etching. The next processing step is the etching step and like development, it is extremely critical. Again both liquid and plasma etching techniques are employed, with the latter rapidly becoming the technique of choice for high resolution lithographic processing. A detailed discussion of plasma etching is given in Chapter 6 of this book and will not be further presented here.

Liquid etching is controlled by the choice of etching reagent, temperature, etching time, degree of agitation, and the age of the etch bath. Liquid etching nearly always results in an undercut, isotropically-etched profile which severely limits its utility for etching thick films. The isotropic nature of this etch makes it important to control precisely the etching end point, and considerable effort is required to achieve reproducible etching times from wafer batch to wafer batch.

4.3.j Stripping. The final step in the lithographic process is resist stripping. This step is, for the most part, non-critical and may be accomplished by either liquid or oxidizing plasma processes. It is important that the stripping procedure not alter the underlying thin film or semiconductor device or induce indesirable contamination that will be deleterious to subsequent lithograhic steps.

Literature Cited

1. Bowden, M. J. *CRC Critical Reviews in Solid State Science*, 1979, *8*, p 223;
2. Thompson, L. F. *Solid State Technol.*, 1976, *17*(7), p 27.
3. Heidenreich, R. D.; Kammlott, G. W. *Procd. of 4th photopolymer conf.*, 1976, Ellenville, N.Y., p 94.
4. Heidenreich, R. D. *J. Appl Phys.*, 1977, *48*, p 1418.
5. Ballantyne, J. P. *J Vac. Sci Technol.*, 1975, *12*, p 1261.
6. Bowden, M. J.; Thompson, L. F.; Ballantye, J. P. *J Vac. Sci Technol.*, 1975, *12*, p 1294.
7. Heidenreich, R. D.; Thompson, L. F.; Feit, E. D.; Melliar-Smith, C. M., *J. Appl. Phys.*, 1973, *44*(9), p 4039.
8. Bowden, M. J. *J. Poly. Sci. Polym. Symp.*, 1974, *49*, p 221.
9. Bowden, M. J.; Thompson, L. F. *J. Appl. Poly. Sci.*, 1973, *17*(10), p 3211.

10. Greeneich, J. S. *J. Appl. Phys.*, 1974, *45*(12), p 5264.
11. Greeneich, J. S. *J. Electrochem. Soc.*; 1974, *121*(12), p 1669.
12. Greeneich, J. S. *J. Electrochem. Soc.*, 1975 *122*(7), p 970.
13. Wolf, E. D.; Ozdemir, F. S.; Perkins W. E.; Coane, P. J. in *11th Symp. on Electron, Ion and Laser Beam Technol.*, Thornley, R. F. *ed*, San Franscio press, San Franscio, CA, 1971, p 331.
14. Engle, P. R. *Proc. Kodak Microelectronics Seminar*, 1975, Monterey, Ca.
15. Bruning, J. H.; Feldman, M.; Kinsel, T. S.; Sittig, E. K.; Townsend, R. L. *IEEE Trans. Electron Dev.* 1975, *ED-22*, p 487.
16. Schoonard, J. W.; Gould, J. D.; Miller, L. A. *IBM Res. Report* 1970, RC3085.
17. Skinner, J. G. *SPIE Semicond. Microlithogr II*, 1977, *100*, p 20; Weiss A. *Semicond. Int* 1982, *5* (9) p. 65.
18. Ballantyne, J. P. "Mask Fabrication by Electron Beam Lithography", Brewer, G. R., ed., Academic Press, New York, 1980, pp 259-307.
19. Price, J. E. *Proc. IEEE*, 1970, *58*, p 1290.
20. Murphy, B. T. *Proc. IEEE*, 1964, *52*, p 1537.
21. Fogiel, M. "Modern Microelectronics"; Research and Education Ass. press; New York, NY, 1972.
22. Maissel, L. I.; Glang, R. "Handbook of Thin Film Technology"; McGraw-Hill, New York, NY, 1970.
23. Harper, C. A. "Handbook of Materials and Processes for Electronics"; McGraw-Hill, New York, NY, 1970.
24. DeForest, W. S. "Photoresist Materials and Processes"; McGraw-Hill, New York, 1975; pp 63-88.
25. Gunderson, R. C.; Hart, A. W. "Synthetic Lubricants"; Reinhold, New York, NY; 1962 pp 11-25.
26. Burggraaf, P. S. *Semicond Int.* 1981, *4*(7), p 71.
27. Washo, B. D. IBM J. Res. Dev., 1977, *21*(2), p 190.
28. Burkman, D. *Semicond. Int.* 1981, *4*(7), p 103.
29. Wolfgang, E. "Microcircuit Engineering", 1980 Nixon, W. C.; Ahmed, H. ed.; Cambridge univ. Press, Cambridge, England, pp 409-446.
30. Craver, J. K.; Tess, R. W. "Applied Polymer Science"; organic Coatings and Plastics chemistry Division of the ACS; Washington, D. C. 1975.
31. Billmeyer, F. W. "Textbook of Polymer Science"; Interscience Pub.: New York, NY, 1978, p 368, 438, 494.
32. Lai, J. H. *Polymer Eng. and Sci.* 1979, *19*(15), p 1117.
33. Damon, G. F. *Proceedings of the 2nd Kodak Seminar on Microminiaturization*, 1967, Rochester, NY.
34. Brandrup, J.; Immergut, E. H. "Polymer Handbook"; Interscience Publishers: New York, NY, 1980.

35. Dean, J. A. "Lange's Handbook of Chemistry"; McGraw-Hill, New York, NY, 1980.
36. Bovey, F. A.; Winslow, F. H. "Macromolecules - An Introduction to Polymer Science"; Academic Press, New York, NY, 1979, pp 273-316.
37. Crank, J.; Park, G. S. "Diffusion in Polymers"; Academic Press, New York, NY, 1968.
38. Wendlandt, W. W. "Thermal Methods of Analysis"; Interscience Publishers, New York, NY, 1964.
39. Bowden, M. J.; Thompson, L. F.; Robinson, W.; Biolsi, M. *Macromolecules*, 1982, *15*, p 1417.
40. Bowden, M. J.; Allara, D. L.; Vroom, W. I.; Frackoviak, J.; Kelley, L. C.; Falcone, D. R. *Org. Coatings and Appl. Poly. Sci. Proceeding*, 1983, *48*, p 161.
41. Bowden, M. J.; Thompson, L. F.; Fahrenholtz, S. R.; Doerries, E. M. *J. Electrochem. Soc.* 1981, *128*(6), p 1304.
42. Thompson, L. F. "Proceedings of the Tutorial symposium on Semiconductor Technology" ed. Doane, D. A.; Fraser, D. B.; Hess, D. W. 1982, *82*(5), p 91.
43. Ohnishi, Y.; Itoh, M.; Mizuno, K.; Gokan, H.; Fujiwara, S. *J. Vac. Sci. Technol.* 1981, *19*(4), p 1141.
44. Taylor, G. N.; Wolf, T. M.; Goldrick, M. R. *J. Electrochem. Soc.* 1981, *128*, p 361.
45. Taylor, G. N.; Wolf, T. M. *Polym. Eng. and Sci.* 1980, *20*, P1087.
46. Taylor, G. N.; Wolf, T. M. *J. Electrochem.* Soc. 1980, *127*, P2665.
47. Goodner, W. R.; Wood, T. E.; Hughes, H. G.; Smith, J. N. *Proc. Kodak Microelect. Sem.* 1979, P 51.
48. Taylor, G. N.; Wolf, T. M. *J. Vac. Sci. Technol.* 1981, *19*(14), p 872.
49. Isaacson, M.; Murray, A. *J. Vac. Sci. Technol.* 1981, *19*(4), p 1117.
50. Tsuda, M.; Oikawa, S.; Kanai, W.; Hashimoto, K.; Yokota, A.; Nuino, K.; Hijikata, I.; Yehara, A.; Nakane, H. *J. Vac. Sci. Technol.* 1981, *19*(4), p 1351.
51. Ueberreiter, K. "Diffusion in polymers"; Crank, J.; Park, G. S. eds. Academic Press, New York, NY, 1968, p 220.
52. Berry, G. C.; Fox, T. G. *Adv. in polym. Sci.* 1968, *5*, p 261.
53. Ueberreiter, K.; Kirchner, P. *Makromol. Chem.* 1965, *87* p 32.
54. Asmussen, F.; Ueberreiter, K. *J. Polymer Sci.* 1962, *57*, p 199.
55. Tu, Y.O.; Ouano, A. C. *IBM Journal* of *Res. and Dev.* 1977, *21*(2), p 131.

RECEIVED February 8, 1983

Plasma Etching

J. A. MUCHA

Bell Laboratories, Murray Hill, NJ 07974

D. W. HESS

Department of Chemical Engineering, University of California
Berkeley, CA 94720

0097-6156/83/0219-0215$16.65/0
© 1983 American Chemical Society

5.1 Introduction

The increase in complexity and the decrease in cost of solid state devices and integrated circuits over the past twenty years has been phenomenal. In large part, such trends have been made possible by a continual reduction in the minimum feature size of individual circuit elements. Currently available exposure tools and highly sensitive resist materials allow the generation of submicron pattern sizes. However, reproducible and controllable transfer (etching) of patterns in the 1-2 μm range is difficult if not impossible using conventional liquid etching. As a result, considerable interest in "dry" etching techniques has developed. Specifically, plasma-assisted etch processes are attractive because of their unique capabilities. In this chapter, we discuss the properties of rf glow discharges (plasmas) as they relate to thin-film etching. Materials commonly used in integrated circuit (IC) fabrication are treated individually. Process considerations such as etch profiles, process monitoring, and safety aspects are described.

5.1.a. Need for Dry Etching. Until recently, liquid etching techniques (*1*) have been the mainstay of pattern delineation methods. Their widespread use arises primarily from two considerations. First, the technology involved in liquid etching is well established, despite the fact that the exact chemistry occurring is often ill-defined. Second, the selectivity (ratio of the etch rate of the film being etched to the etch rate of the underlying film or substrate) is generally infinite with typical liquid etchant systems.

Unfortunately, solution etching presents several critical problems for micron and submicron geometries. Because of the acid environments of most etchant solutions, photoresists can lose adhesion, thereby altering patterning dimensions and preventing linewidth control (*2*). In addition, as etching proceeds downward, it proceeds laterally at essentially an equal rate. This process undercuts the mask and generates an isotropic etch profile as shown in Figure 1. Because the film thickness and the etch rate are often non-uniform, a certain degree of "overetching" is required. Overetching results in an alteration in pattern size and thus process control. Naturally, if the film thickness is small relative to the minimum pattern dimension, undercutting is inconsequential. However, when the film thickness is comparable to the lateral film dimension, as is the case for current and future devices, undercutting can be intolerable.

Finally, as device geometries decrease, spacings between stripes of resist also decrease. With micron and submicron patterns, the surface tension of etch solutions can cause the liquid to bridge the space between two resist stripes. Etching of the underlying film is thereby precluded.

Plasma etching has demonstrated viable solutions to essentially all the problems encountered with liquid etching. Adhesion does not appear to be critical with dry etching techniques. In addition, undercutting can be

ISOTROPIC VERSUS ANISOTROPIC ETCHING

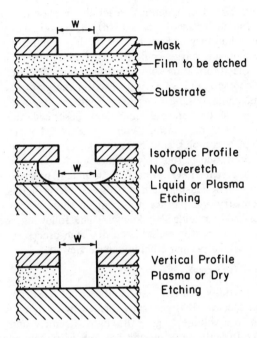

Figure 1. Cross sections of films etched with liquid or plasma etchants. The isotropic profile represents no overetch, and can be generated with liquid or plasma etch techniques. The anisotropic profile requires plasma or dry etch processes.

controlled by varying the plasma chemistry, the gas pressure, and the electrode potentials. As a result, considerable effort has been expanded recently to replace the firmly established wet etch processes with plasma or dry etch technologies.

From an IC manufacturing standpoint, two additional considerations fuel the drive toward dry etch processes. Relatively large volumes of dangerous acids and solvents must be handled and ultimately recycled or disposed of with wet etching or resist stripping techniques. Dry etching or resist stripping operations use comparatively small amounts of chemicals.

However, these materials may also be quite corrosive and toxic, as will be discussed later. Finally, unlike wet processing methods, plasma techniques are conducive to fully automated processes.

5.1.b. Dry Etching History. Chemical reactions in glow discharges or plasmas have been studied for many years (*3-6*). However, many of these studies were merely laboratory curiosities. In the late 1960's, it was suggested that an oxygen glow discharge could be used to remove photoresist films rapidly during the fabrication of microelectronic devices (*7*). Shortly thereafter, fluorine-containing gases were used in the glow discharge etching of silicon and silicon-containing films (*8*). Such results, combined with the obvious advantages of dry process technology described above, stimulated the commercialization of plasma etch systems. Unfortunately, glow discharges are highly complex entities, and control of the etch process soon became a major limitation of this new technology.

The past five years have witnessed significant progress in the understanding and control of plasma etch processes. Nevertheless, much of the fundamental chemistry and physics occurring in these reactive atmospheres remains ill-understood, or indeed unknown. The following sections assimilate the information currently available on dry etch processes, and present a framework within which plasma etching can be viewed.

5.2 Glow Discharges (Plasmas)

A plasma is a partially ionized gas composed of electrons, ions, and a variety of neutral species. It is generally formed by applying a large electric or magnetic field to a volume of gas, although extremely high temperatures can also be used. In addition, a plasma has approximately equal concentrations of positive and negative charge carriers. Many types of plasmas exist (*9*); they differ primarily in electron concentration (n_e), and in average electron energy, kT_e, where k is Boltzmann's constant, and T_e is the electron temperature. For instance, high pressure arcs are characterized by $n_e \approx 10^{16}$ cm^{-3} and $kT_e \approx 0.5$ eV, while in interplanetary space, $n_e \approx 10$ cm^{-3} and $kT_e \approx 25$ eV.

A quantity which is useful in characterizing the average electron energy in a plasma is the ratio of the electric field strength to the pressure, E/P (*9*). As the electric field increases, free electrons, whose velocities increase by the action of the field, gain energy. However, since they lose this energy by collisional processes, an increase in pressure, which decreases the mean free path of the electrons, decreases the electron energy.

In thin film processes for microelectronic applications, we deal almost exclusively with glow discharges. These plasmas are characterized by pressures in the range of 50 mTorr to 5 Torr, electron densities between 10^9 and 10^{12} cm^{-3}, and average electron energies between 1 and 10 eV (such energies correspond to temperatures of $10^4 - 10^5$ °K). Further, glow discharges are termed nonequilibrium plasmas since the electron temperature is greater than the ion or gas temperature (T_i); typically, $T_e/T_i > 10$. Finally, glow discharge plasmas are only weakly ionized; i.e., the ratio of n_e to the neutral species concentration (n_N) is in the range 10^{-6} to 10^{-4}.

The above characteristics impart glow discharge plasmas with unique and useful properties. Energetic electrons are capable of breaking chemical bonds via collisions with gas molecules. In this way, highly reactive chemical species are produced for etching (or deposition) reactions. Since the gas temperature remains near ambient, temperature sensitive materials can be processed in this environment. Also, the surface chemistry occurring in glow discharges is generally modified by the impingement of ions and electrons onto the film being etched. The combination of these physical processes with the strictly chemical reactions (due to atoms or molecules) results in etch rates and etch profiles unattainable with either process individually. Thus the role of the plasma in etching processes is twofold (*10*): create chemically reactive species, and provide energetic radiation (principally ions and electrons, but also photons) for alteration of surface chemistry.

5.2.a. Physical and Electrical Aspects. The energy of ions and electrons striking surfaces in a glow discharge is determined by the potentials established within the reaction chamber. Since etching (and deposition) of thin films is generally carried out in an rf plasma generated by capacitively coupling the rf field to the plasma (*11*), this type of configuration will be the only one discussed in the present section. The potentials of importance in rf glow discharge systems (*12,13*) are the plasma potential (the potential of the glow region), the floating potential (the potential assumed by a surface within the plasma which is not externally biased or grounded, and thus draws no *net* current) and the potential of the powered or externally biased electrode. When the plasma contacts a surface, that surface, even if grounded, is always at a negative potential with respect to the plasma (*12,14,15*). Therefore, positive ion bombardment occurs. The energy of the bombarding ions is established by the difference in potential between the plasma and the surface which the ion strikes. Because these potentials may range from a few volts to a thousand volts, surface bonds can be broken, and in certain instances sputtering of film or electrode material may occur (*12*).

In addition to sputtering and surface chemistry modification, exposure of materials to energetic radiation can result in radiation damage (*16*). This damage originates from several sources. Positive ions can cause implantation or displacement damage, while electrons, x-rays, or UV photons can result in ionization. The defects thereby created can serve as trapping sites for electrons or holes, resulting in an alteration of the electrical properties of the materials. Although ionization may be eliminated by low temperature ($< 500°C$) heat treatments, implantation or displacement damage may require temperatures above $900°C$ for complete elimination.

The reason for the different potentials within a plasma system can be seen easily by considering electron and ion mobilities (14). Imagine applying an rf field between two plates (electrodes) positioned within a low-pressure gas. On the first half-cycle of the field, one electrode is negative and attracts positive ions; the other electrode is positive, and attracts electrons. Because of the frequencies used (50 kHz to 40 MHz) and because the mobility of electrons is considerably greater than that of positive ions, the flux (current) of electrons is much larger than the positive ion flux. This causes a depletion of electrons in the plasma, and thus a positive plasma potential. On the second half cycle, a large flux of electrons flows to the electrode that previously received the small flux of ions. Further, since plasma etching systems generally have a dielectric coating on the electrodes, and/or have a series (blocking) capacitor between the power supply and the electrode, no charge can be passed. Therefore, on each subsequent half-cycle negative charge continues to build on the electrodes (and on other surfaces in contact with the plasma) so that electrons are repelled and positive ions attracted to the surface. Of course, this transient situation ceases when a sufficient negative bias is achieved on the electrodes such that the fluxes of electrons and positive ions striking these surfaces are equal. At this point, time-average (positive) plasma and (negative) electrode potentials are established.

The plasma potential is essentially uniform throughout the observed glow volume in an rf discharge. Between the glow and the electrode is a narrow region wherein a change from the plasma potential to the electrode (or surface) potential occurs. This region is called a sheath or a dark space, and is shown schematically in Figure 2. Positive ions that reach the edge of the glow region are accelerated across the potential drop and strike the electrode or substrate surface.

Because of the series capacitor and/or the dielectric coating of the electrodes, the negative potentials established on the two electrodes in a plasma system may not be the same. For instance, the ratio of the voltages on the electrodes has been shown to be dependent upon the relative electrode areas (17). The (theoretical) dependence is given by Equation 1, where V is the voltage and A is the area (16).

$$\frac{V_1}{V_2} = \left(\frac{A_2}{A_1}\right)^4 \tag{1}$$

If V_1 is the voltage on the powered electrode, and V_2 is the voltage on the grounded electrode, the voltage ratio is the inverse ratio of the electrode areas raised to the fourth power. However, for typical etch systems, the

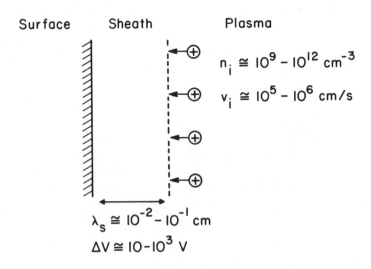

Figure 2. Schematic diagram of a solid surface exposed to a plasma. Typical values for important parameters are indicated. The symbols λ_s, ΔV, n_i, and v_i represent the sheath thickness, sheath potential, ion concentration, and ion velocity, respectively.

exponent of the area ratio is generally less than 4 (12). Although the actual electrodes in a plasma reactor often have the same area, A_2 is the *grounded* electrode area i.e., the area of all grounded surfaces in contact with the plasma. Since this usually includes the chamber walls, the area ratio can be quite large. Because of such considerations, the average potential distribution in a typical commercial plasma reactor with two parallel electrodes immersed in the plasma is similar to that shown in Figure 3 (12). In this case, the energy of ions striking the powered electrode (or substrates on this electrode) will be higher than that of ions reaching the grounded electrode.

   In addition to the electrode area ratio, other plasma parameters can affect the electrical characteristics of the discharge. For instance, variation of the rf power input will alter plasma and electrode potentials, thereby changing ion energies. Also, rf frequency affects the kinetic energy of ions that strike surfaces in contact with the plasma. This can be readily understood by considering the behavior of an ion in the plasma experiencing an oscillating plasma potential due to applied rf voltages (18,19). Depending upon the ion mobility, some frequency exists above which the ion can no

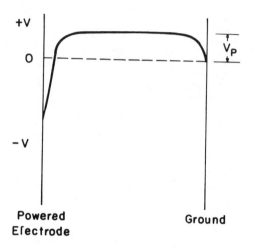

Figure 3. Potential distribution in a parallel plate plasma etcher with grounded surface area larger than powered electrode area. (Reproduced with permission from Ref. 12.)

longer follow the alternating voltage. Therefore, the ion cannot traverse the sheath in one half-cycle. Above this frequency, ions experience an accelerating field (difference between the plasma and electrode potentials divided by the sheath thickness) that is an average over a number of half-cycles. At lower frequencies, where the ions can respond directly to the oscillating field they are accelerated by instantaneous fields. Thus, the ions can attain the maximum energy corresponding to the maximum instantaneous field across the sheath. As a result, for a constant sheath potential, ion bombardment energies are higher at lower frequencies.

5.2.b. Chemical Aspects. Ideally, plasma etching can be broken down into six primary steps, as shown in Figure 4. If any of these individual processes does not occur, the overall etch cycle terminates. First, reactive species (atoms, molecules, or ions) must be generated by electron/molecule collisions. This step is vital in plasma etching, because many of the gases used to etch thin film materials do not react spontaneously with the film. For instance, carbon tetrafluoride (CF_4) does not etch silicon. However, when CF_4 is dissociated via electron collision to form fluorine atoms, etching of silicon occurs readily.

Etchant species (for example, fluorine atoms) diffuse to the surface of the material and adsorb onto a surface site. It has been suggested (*20*) that free radicals have fairly large sticking coefficients compared with relatively inert molecules such as CF_4, so adsorption occurs easily. In addition, it is generally assumed (*20*) that a free radical will chemisorb and react with a solid surface. Further, surface diffusion of the adsorbed species or of the product molecule can occur.

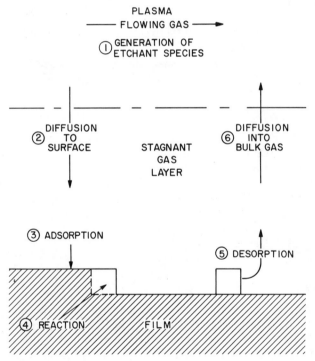

Figure 4. *Primary processes occurring in a plasma etch process.*

Product desorption is a crucial step in the etch process. A free radical can react rapidly with a solid surface, but unless the product species has a reasonable vapor pressure so that desorption occurs, no etching takes place. For instance, when an aluminum surface is exposed to fluorine atoms, the atoms adsorb and react to form AlF_3. However, the vapor pressure of AlF_3 is ~1 torr at 1240°C; thus etching is precluded at ambient temperatures.

The chemical reactions taking place in glow discharges are often extraordinarily complex. However, two general types of chemical processes can be categorized: homogeneous gas-phase collisions and heterogeneous surface interactions. In order to completely understand and characterize plasma etching, one must understand the fundamental principles of both processes.

Homogeneous gas-phase collisions represent the manner in which reactive free radicals, metastable species, and/or ions are generated. As shown in Table I, electron impact can result in a number of different reactions (21-23).

Table I. Examples of Electron Impact Reactions (21)

1. **Excitation (rotational, vibrational, electronic)**
   $$e + A_2 \rightarrow A_2^* + e$$

2. **Ionization**
   $$e + A_2 \rightarrow A_2^+ + 2e$$

3. **Dissociative Ionization**
   $$e + A_2 \rightarrow A^+ + A + 2e$$

4. **Dissociation**
   $$e + A_2 \rightarrow 2A + e$$

5. **Dissociative Attachment**
   $$e + A_2 \rightarrow \begin{cases} A^- + A^+ + e \\ A^- + A \end{cases}$$

Due to the electronegative character of many of the etch gases currently used (e.g., O_2, CF_4, CHF_3, CF_3Cl, CCl_4, BCl_3, etc.), electron attachment often takes place, thereby generating negative as well as positive ions in the plasma atmosphere. Although these negative ions affect the plasma energetics, they probably have little if any effect on surface reactions, because they are repelled by the negative electrode potential.

Electron impact reactions occur at a rate determined by the concentrations of both electrons (n_e) and a particular reactant (N) species (*21*)

$$R = kn_e N \tag{2}$$

The proportionality constant k is the rate coefficient, which is generally expressed by the following equation (*21*)

$$k = \int_0^\infty (\frac{\epsilon}{2m})^{1/2} \sigma(\epsilon) f(\epsilon) d\epsilon \tag{3}$$

In Equation 3, ϵ and m are the impinging electron energy and mass, $\sigma(\epsilon)$ is the reaction cross section, and $f(\epsilon)$ is the electron energy distribution function. Of course, if an accurate expression for $f(\epsilon)$ and if electron collision cross sections for the various gas phase species present are known, k can be calculated. Unfortunately, such information is generally unavailable for the types of molecules used in plasma etching.

It should be noted that a Maxwellian form of $f(\epsilon)$ is a reasonable approximation to the actual distribution at low electron energies. This observation is indicated in Figure 5 (*24*). However, the first ionization potential of most atoms and molecules is above ~8 eV. Thus, many of the important homogeneous processes that occur in glow discharges, such as ionization, take place as a result of high energy electrons in the "tail" of the distribution. These electrons are precisely the ones that are not adequately described by a Maxwellian distribution function.

A second type of gas phase collision is that occurring between the various (heavy) species generated by electron impact reactions, as well as between these species and the unreacted gas-phase molecules (*25,26*). Again, dissociation and ionization processes occur, but in addition, recombination and molecular rearrangements are prevalent. Similar rate expressions to that of Equation 2 can be written for these collisions (*21*). In this case, the concentration of each chemical species, along with the collision cross section, and the species' energy distribution function must be known if k is to be calculated. Clearly, much of this information is presently unknown.

A wide spectrum of heterogeneous processes can occur at solid surfaces exposed to a glow discharge (*3,26,27*). The primary processes of interest in plasma etching (and in plasma deposition) are summarized in Table II (*20*). These interactions result from the bombardment of surfaces by particles. Although vacuum UV photons and soft x-rays present in the plasma are sufficiently energetic to break chemical bonds, electron and particularly ion bombardment has been found to be the most effective method of promoting surface reactions (*28*).

The enhancement of etch rates due to particle bombardment can be simply envisioned by referring to Figure 4. Steps 3-5 are heterogeneous

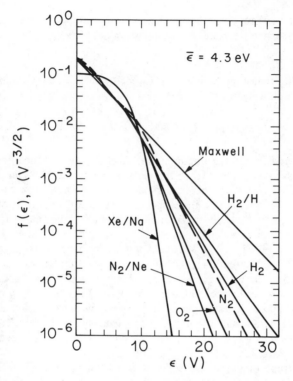

Figure 5.   Electron energy distribution functions for various gases and gas mixtures. (Reproduced with permission from Ref. 24.)

Table II. Elementary Processes at Solid Surfaces Exposed to a Glow Discharge (20)

1. **Ion-Surface Interactions**
    a.   Neutralization and secondary electron emission
    b.   Sputtering
    c.   Ion-induced chemistry

2. **Electron-Surface Interactions**
    a.   Secondary-electron emission
    b.   Electron-induced chemistry

3. **Radical- or Atom-Surface Interactions**
    a.   Surface etching
    b.   Film deposition

processes whose kinetics are temperature dependent. However, temperature is merely one method of increasing the energy that surface bonds possess. Particle bombardment is another means by which energy is imparted to a surface. Specifically, ion (or electron) bombardment can break surface bonds, thereby creating crystal damage and adsorption sites (29) as well as assisting product desorption (30). Also, chemical reactions on the solid surface can be promoted by such bombardment (28). Which of these steps is primarily responsible for enhanced etch rates is not yet clear. Nevertheless, particle bombardment does promote etch processes, as demonstrated in the beam experiments described by Figure 6. In this study, a beam of XeF_2 molecules and a beam of argon ions were directed at a silicon film (31). Xenon difluoride was used because it afforded a method of "depositing" fluorine atoms onto the silicon surface. Measurement of the silicon etch rate as a function of XeF_2 or Ar^+ exposure allowed a comparison of chemical and physical etch processes. When only XeF_2 contacted the silicon surface, a small (~5 Å/min) purely chemical etch reaction was observed.

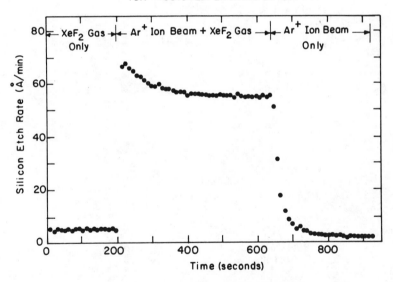

Figure 6.   Example of ion-assisted gas-surface chemistry in the etching of silicon with XeF_2. The XeF_2 flow is 2×10^{15} moles/sec and the argon energy and current are 450 eV and 2.5 μA, respectively. (Reproduced with permission from Ref. 31.)

Likewise, when Ar^+ impinged on the surface, pure sputtering (~ 2 Å/min) was noted. However, when the beams were simultaneously directed at the silicon surface, a relatively large (~ 55 Å/min.) etch rate was observed; the measured rate was approximately an order of magnitude greater than the sum of the chemical and physical components. Obviously, synergistic effects due to ion bombardment are crucial to this chemical etch process. Unfortunately, the exact nature of these effects is at present undefined.

5.2.c. Parameter Control. At present, the most serious impediment to routine use of plasma etching is the large number of parameters that affect the process. As noted, both gas phase considerations and plasma-surface interactions must be controlled. The problem is illustrated in Figure 7 (*32*). Naturally, if the basic plasma parameters (N, n_e, $f(\epsilon)$, τ) could be controlled, one could define the gas phase chemistry. Even in this case, the nature of the particle-surface interactions is unknown.

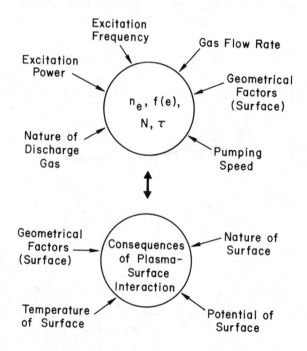

Figure 7. Representation of the parameter problem in plasma processes. The symbols n_e, $f(\epsilon)$, N, and τ are electron density, electron energy distribution, gas density, and residence time for molecules in the plasma volume, respectively. (Reproduced with permission from Ref. 32.)

Although many macroscopic parameters (i.e., power, frequency, discharge gas, etc.) can be varied, the precise manner in which a change in any of these variables affects the basic plasma parameters is unknown. Indeed, a change in a macroscopic parameter generally results in a change in two or more basic plasma parameters, as well as surface potential and surface temperature. For instance, rf power determines the current and voltage between the electrodes in a plasma system. A variation in rf frequency changes the number and the energy of ions that can follow the changing field, and thus affects the bombardment flux and energy. The gas flow rate, the pump speed, and the pressure are all interrelated, because two ways of altering the gas pressure can be envisioned. The pump speed can be held constant and the gas flow rate varied; alternatively, the gas flow rate may be held constant and the pump speed varied (by throttling the pump). These two methods of pressure variation can result in different pressure dependencies of the etch rate due to the different residence time for the two processes. The specific etchant gas and the surface temperature (not necessarily the same as the electrode temperature) are vital parameters because of the dependence of the etch process on reactive species concentrations and on the observation that many of the etch rates follow an Arrhenius rate expression. Electrode (and chamber) material can alter the chemistry occurring in glow discharges because of chemical reactions (atom adsorption, recombination, etc.) on the electrode surface or with the electrode itself. Finally, the electrode potential and, as indicated by Equation 1, the reactor configuration determine the energy of ions and electrons striking the surfaces in contact with the plasma.

5.3 Plasma Etching Systems

A plasma etching system has several basic components: a vacuum chamber and a pumping system to maintain reduced pressures; a power supply to create the glow discharge; and gas or vapor handling capabilities to meter and control the flow of reactant gases. Detailed assembly of such a system is subject to wide variations depending upon the parameter space, parameter control, and application desired. For the most part, however, this multitude of possibilities can be discussed in terms of three reactor configurations: barrel or volume-loaded; parallel plate or planar; and downstream.

The earliest and simplest plasma apparatus constructed for etching was the barrel reactor, shown schematically in Figure 8a. This configuration generally uses a cylindrical chamber with rf power applied to external coils or to external electrodes. The wafers are placed in a holder or "boat" within the chamber. As a result of concerns regarding temperature uniformity along the length of the holder and radiation damage, a perforated cylindrical "etch tunnel" is often used in the barrel reactor. This metal cylinder confines the glow region to the annulus between the etch tun-

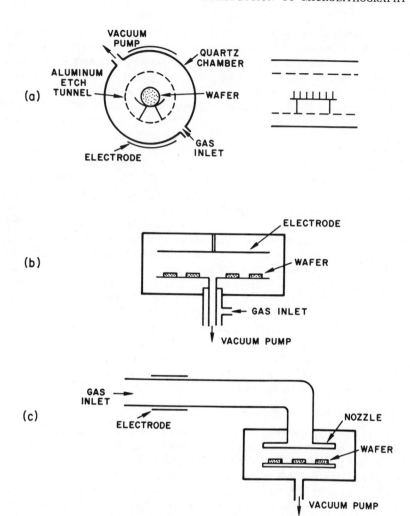

Figure 8.   Configurations for plasma etch reactors.   (a) barrel or volume loaded; (b) parallel plate or surface loaded; (c) downstream etcher.

nel and the chamber wall, and shields the substrates from energetic ions and electrons in the plasma. Because the wafers are shielded by the etch tunnel, they are subject to little, if any, bombardment. In this case the etch reaction is almost purely chemical, resulting in an isotropic etch profile. Therefore, barrel systems are most often used for resist stripping or in "noncritical" etching steps, where undercutting can be tolerated.

Since very large scale integration (VLSI) requires nearly vertical etch profiles, the use of planar or parallel plate reactors is favored for many etch processes. In this configuration (Figure 8b), which is similar to that utilized in rf sputtering, the wafers lie on an electrode within the plasma, causing them to be exposed to energetic radiation. As discussed earlier, such geometries result in ion bombardment, and thus in both physical and chemical components to the etch process. The physical component can impart directionality to the etch process.

The planar system can be operated in two modes. When the wafers reside on the grounded electrode while the opposite electrode is powered, the process is generally termed plasma etching. When substrates are positioned on the powered electrode, a reactive ion etching (RIE) or reactive sputter etching configuration results (*33,34*). The RIE arrangement generally causes the wafers to be subjected to higher energy ion bombardment than in the plasma etching mode because of the potentials established on the grounded versus the powered electrode (*see* Section 5.2.a.). Another reason for higher energy ions in RIE is the lower operating pressure. Although no exact demarcation in pressure exists, typically RIE is performed at pressures below 100 mTorr, while plasma etching is carried out at pressures above this value. The above ion energy considerations often lead to a greater possibility of radiation damage in RIE than in plasma etching. However, it should be noted that ion bombardment energies above 200 eV are possible even at high pressures (> 100 mTorr). The controlling factor is the difference in potential between the plasma and the electrodes.

Before leaving the subject of planar etching systems, a comment should be made concerning the term reactive ion etching. This term suggests that in a plasma system, reactive ions are the important species in the etch reaction. Clearly, reactive ions can chemically etch materials (*see* Section 5.7.b.). However, because the concentration of neutral species greatly exceeds that of positive ions in the plasma and considering the high etch rates obtained, one must conclude that free radicals are the primary etchant species. The term ion-assisted etching (*11,31*) therefore appears to be a more accurate and descriptive term than reactive ion etching.

In principle, downstream etching configurations (Figure 8c) offer several advantages over barrel and planar systems. For instance, precise temperature control can be difficult in either a barrel or a planar etcher, and radiation damage is a concern when the wafers are immersed in the plasma. Downstream etchers have used microwave radiation to generate long-lived chemical species which are then transported to wafers located downstream from the plasma (*35*). Temperature control and radiation damage problems are thereby minimized or eliminated. However, in instances in which a material (i.e., Al_2O_3) requires energetic radiation to be etched at an appreciable rate or in which isotropic etching cannot be tolerated, downstream systems may be inappropriate.

5.4 Etching Considerations

5.4.a. General Considerations. The simple concept of choosing a gas that can be dissociated in a discharge to form reactive species that can lead to a volatile etch product has been a significant driving force behind the development of etchants for plasma processing. As noted earlier, the situation becomes quite complex when one considers the number of physical and chemical parameters that must be controlled in applying this concept. Although the parameter space has been given in terms of variables that are controlled or measured by the operator, the characterization of the resulting etch process becomes the "bottom line" for plasma processing. Thus, plasma etching techniques may be classified according to rate, selectivity, anisotropy, uniformity, and surface texture. Etch rates determine the time for processing in a given plasma environment, while the ratio of rates for two different materials immersed in the same plasma determines the selectivity for removing one film without significantly etching an underlying one. Preferential etching in a direction normal to the film surface is termed anisotropy (*see* Section 5.5.c.) and is an important etch characteristic for high-density, fine-line patterning. The term "loading" is applied to plasmas in which a measurable depletion of the active etchant is caused by consumption in the etch process. Texture describes the quality of the etch. Some plasmas leave a smooth substrate surface, while others tend to roughen or texture an initially smooth surface.

High etch rates and selectivity can be achieved by judicious selection of feed gases to a plasma reactor. The atomic and radical species formed by electron impact dissociation depend largely on feed gas composition, and the intrinsic etch rates measured in the absence of a plasma (i.e., downstream etching) provide a useful indicator of chemical selectivity in the presence of a plasma. For example, the ratio of (100) silicon (*34*) to thermal oxide (SiO_2) (*37*) etching by F atoms is 41:1 at room temperature. As etch rates generally follow an Arrhenius type dependence on substrate temperature,

$$\text{Etch rate} \propto \exp(-E_a/kT), \tag{4}$$

and the effective activation energy E_a is material dependent, selectivity varies exponentially with temperature. Since activation energies for chemical processes are in the range 0.1 - 2.0 eV, selectivity, which is based on ratios of E_a's, becomes a very sensitive function of temperature. Figure 9 shows this dependence for the $Si/SiO_2/F$ system, and similar variations apply to other materials.

Ion energy and flux represent another important aspect in achieving and controlling selectivity. Because ion bombardment generally imparts different degrees of damage to different materials, etch selectivity can often be varied and even reversed from that obtained via chemical means. In

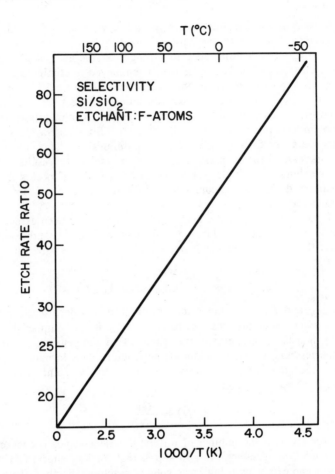

Figure 9. Silicon to SiO₂ etch rate ratio for atomic fluorine etching as a function of substrate temperature. Constructed from etch rate expressions in Refs. 36 and 37.

addition, no temperature dependence is expected when ion-enhanced etching dominates. Under these conditions, however, etch rates should be interpreted only qualitatively since the electrical characteristics of the plasma reactor are rate controlling.

Most isotropic etchants generally exhibit a pronounced loading effect resulting from depletion of the active etchant species by reaction with the film. The degree of loading depends strongly on the area of film exposed. For research studies small samples can be employed to eliminate this effect; however, on a production scale the degree of loading can severely reduce etch rates. In extreme cases, etchant depletion can be so severe that polymer deposition occurs instead of etching. The most common practical loading problem occurs at the endpoint when the film/substrate interface is exposed. Here, the area of etchable surface diminishes, the concentration of etchant increases, and a marked increase in etch rate results. Because chemical etching is isotropic, rapid lateral "undercutting" occurs in regions where pattern delineation is complete and the linewidth control imparted by the mask is reduced.

An analysis (38) of the loading effect, which has been extended (39) to include multiple etchant loading and other etchant loss processes, indicates that the etch rate (R) for N wafers each of area A is given by

$$R(N) = \frac{(k_{etch}/k_{loss})G}{1 + (k_{etch}\ \rho\ NA/k_{loss}V)} \tag{5}$$

where k_{etch} and k_{loss} are the rate constants (first order) for etching and etchant loss in an "empty" reactor, respectively; ρ is the "number density" of substrate molecules per wafer in the reactor; G is the generation rate of the active species; and V is the volume of the reactor. This formalism indicates that as the number of wafers increases to a point where $k_{etch}\ \rho\ NA/k_{loss}\ V \gg 1$ the etch rate

$$R(N) = \frac{GV}{\rho NA} \tag{6}$$

varies inversely with the number of wafers; this dependence is characteristic of a marked loading effect. It also shows that loading effects can be controlled to some extent by using large volume reactors. Further, these effects can be eliminated when homogeneous or heterogeneous etchant loss dominates ($k_{loss} \gg k_{etch}$). When ion bombardment controls the reaction, Equation 5 is inapplicable since the etch rate is determined by bombardment rate rather than etchant supply.

Other transport limitations, such as diffusion-controlled reactions, can lead to localized depletion of etchant, which results in a number of observable etch effects. The size and density of features can influence the etch rate at different locations on a single wafer, thus producing "pattern sensitivity." Depletion across a wafer produces a "bulls eye" effect, while depletion across a reactor is indicated by the fact that the leading wafer edge etches faster than the trailing edge. Similar effects are noted when product removal is transport-limited. Most of these effects can be reduced by judicious choice of pressure and flow rates and by precisely controlling the termination of the etch process using suitable end-point detection (Section 5.6). Pressure and flow interact by determining the residence time relative to diffusion, convection, and reaction rates. Residence times that are short compared with reaction times will reduce diffusion limitations, while long residence times will enhance them.

5.4.b. Chemical Models.   As discussed above, an extensive parameter space is associated with plasma techniques. Therefore, if the development of etch processes is to proceed efficiently, some means of data assimilation and prediction must be available. Two general schemes have been proposed to organize chemical and physical information on plasma etching. Both have dealt primarily with carbon-containing gases, but with slight modifications, they can be easily applied to other etchants. Conceptually, the two models are similar, although they emphasize different aspects of plasma etching.

The fluorine to carbon ratio (F/C) of the "active species" can be used (*40*) to explain observed etch results (Figure 10). This model does not con-

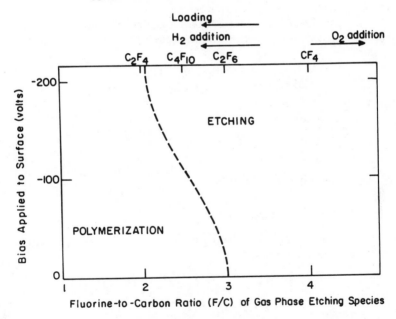

Figure 10. Schematic of the influence of fluorine-to-carbon ratio and electrode bias on etching versus polymerization processes in fluorocarbon plasmas. (Reproduced with permission from Ref. 40.)

sider the specific chemistry occurring in a glow discharge, but rather views the plasma as a ratio of fluorine to carbon species which can react with a (silicon) surface. The generation or elimination of these "active species" by various processes or gas additions then modifies the initial F/C ratio of the inlet gas.

The F/C ratio model accounts for the fact that in carbon-containing gases, etching and polymerization occur simultaneously. The process that dominates depends upon etch gas stoichiometry, reactive gas additions, amount of material to be etched, and electrode potential and upon how these factors affect the F/C ratio. For instance, as described in Figure 10, the F/C ratio of the etchant gas determines whether etching or polymerization is favored. If the primary etchant species for silicon (F atoms) is consumed either by a loading effect or by reaction with hydrogen to form HF, the F/C ratio decreases, thereby enhancing polymerization. However, if oxygen is added to the etchant gas, reaction with carbon-containing fragments to form CO or CO_2 can occur, thus increasing the F/C ratio, and favoring etching. Further, as the negative bias of a surface exposed to the plasma increases at a constant F/C ratio, etching of this surface becomes more important relative to polymerization. Such effects are caused primarily by enhanced energies of the ions striking these surfaces.

In the etchant-unsaturate model described by Equation 7-10 (*41*), specific chemical species derived from electron collisions with etchant gases are considered.

$$e + \text{Halocarbon} \rightarrow \frac{\text{Saturated}}{\text{Radicals}} + \frac{\text{Unsaturated}}{\text{Radicals}} + \text{Atoms} \qquad (7)$$

$$\left\{ \begin{array}{l} \text{Reactive Atoms} \\ \text{Reactive Molecules} \end{array} \right\} + \text{Unsaturates} \rightarrow \text{Saturates} \qquad (8)$$

$$\text{Atoms} + \text{Surfaces} \rightarrow \left\{ \begin{array}{l} \text{Chemisorbed Layer} \\ \text{Volatile Products} \end{array} \right. \qquad (9)$$

$$\text{Unsaturates} + \text{Surfaces} \rightarrow \text{Films} \qquad (10)$$

Application of this model to a CF_4 plasma results in the chemical scheme indicated by Equations 11-14.

$$2e + 2CF_4 \rightarrow CF_3 + CF_2 + 3F + 2e \qquad (11)$$

$$F + CF_2 \rightarrow CF_3 \qquad (12)$$

$$4F + Si \rightarrow SiF_4 \qquad (13)$$

$$n\ CF_2 + \text{Surface} \rightarrow (CF_2)_n \qquad (14)$$

Depending upon the particular precursors generated in the gas phase, etching, recombination, or film formation (i.e., polymerization) can occur. Also, gas-phase oxidant additives (O_2, F_2, Cl_2, etc.) can dissociate and react with unsaturate species. As an example, O_2 can undergo the following reactions in a CF_4 plasma:

$$e + O_2 \rightarrow 2\,O + e \tag{15}$$

$$O + CF_2 \rightarrow COF_{2-x} + x\,F. \tag{16}$$

Mass spectrometer studies of oxidant additions to fluoro- and chlorocarbon gases have demonstrated that the relative reactivity of atoms with unsaturate species in a glow discharge follows the sequence $F \sim O > Cl > Br$ (41). Of course, the most reactive species present will preferentially undergo saturation reactions that reduce polymer formation and that may increase halogen atom concentration. Ultimately, determination of the relative reactivity of the plasma species allows prediction of the primary atomic etchants in a plasma of specific composition.

5.4.c. Silicon-Based Materials.

5.4.c.1. Silicon and Polysilicon. The isotropic etching of silicon (Si) and polycrystalline silicon (Poly-Si) by atomic fluorine (F) is probably the most completely understood of all etch processes, particularly for the cases in which F atoms are produced in discharges of F_2 (36) and CF_4/O_2 (42). Fluorine atoms etch (100) Si at a rate (Å/min) given by (36):

$$R_{F(Si)} = 2.91 \times 10^{12} \, T^{\frac{1}{2}} \, n_F \exp(-0.108 \, eV/kT) \tag{17}$$

where n_F is the F-atom concentration (cm^{-3}), T the temperature (K), and k is Boltzmann's constant. In addition to defining the intrinsic etch rate of F atoms, chemiluminescence arising from the gas phase reactions (36,43)

$$SiF_2 + F \; (F_2) \; \rightarrow SiF_3^*(+F) \tag{18}$$

$$SiF_3^* \rightarrow SiF_3 + h\nu \tag{19}$$

coupled with other spectroscopic evidence (44,45) has provided an understanding of the detailed mechanism of silicon etching. Figure 11 summarizes the surface reactions leading to the gasification of Si to form SiF_2 and SiF_4 as the dominant products. The steady-state surface appears to be a stable "SiF_2-like" (44) structure that must be penetrated by impinging F atoms. This step accounts for the low reaction probability ($\epsilon < 0.01$) (36) at typical substrate temperatures. Reaction I (Figure 11) is rate limiting and proceeds at a rate defined by Equation 17. Since other F-containing feed gases shown in Table III exhibit similar activation energies, it is likely that the same mechanism and etchant (F atoms) are operative in those systems and that variations in etch rate are simply due to variability in F content.

Table III: Etch Rates and Activation Energies
for Silicon Etching in F-Source Plasmas

Plasma	Etch Rate at 100°C (Å/min)	E_a (eV)	Reference
$F \cdot /F_2$[†]	4600	0.108	*36*
CF_4/O_2	4600	0.11	*38*
CF_4/O_2	3000	0.11	*46*
CF_4	300	0.124	*47*
SiF_4/O_2	440	0.11	*46*

[†]etching downstream from r.f. discharge of F_2

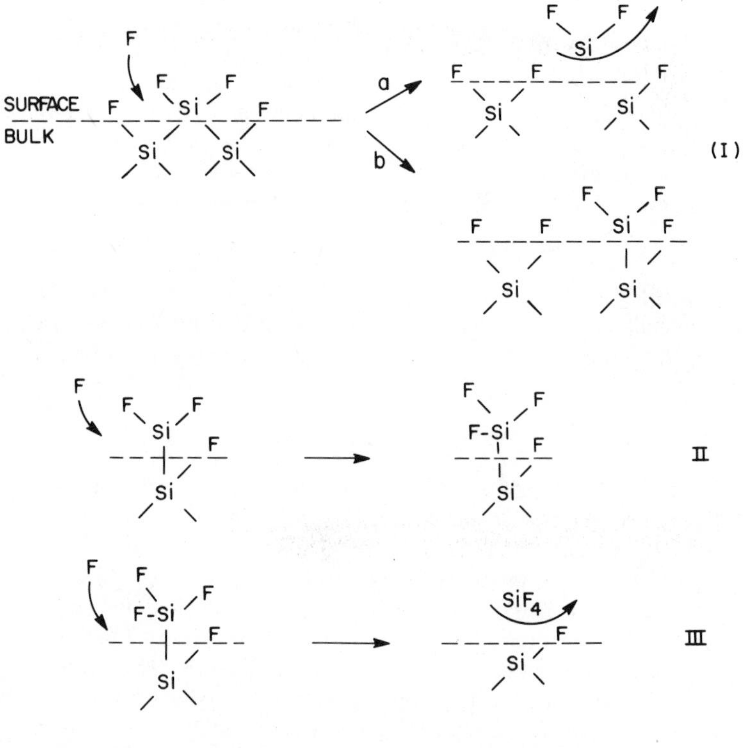

Figure 11.   Mechanism for F-atom reaction with a silicon film leading to the products SiF_2 and SiF_4.

Other F-forming plasmas that have been used successfully in etching Si and Poly-Si include SiF_4, SiF_4/O_2 (*48*), SF_6 (*49,50*), SF_6/O_2 (*50*), NF_3 (*39*), and ClF_3 (*39*). In all cases high selectivity over SiO_2 (10-40:1) and Si_3N_4 (5-10:1) could be achieved.

The use of gases such as CF_4 and SF_6 offer advantages because of their low toxicity; however, the formation of unsaturates such as C_xF_{2x} and S_xF_{2x} in the discharge can scavenge free F atoms, and in unfavorable cases can lead to polymer or residue formation. The role of oxygen in these plasmas has been found to be two-fold. First, in accord with the etchant-unsaturate model, O_2 (and O atoms) reacts with unsaturates to enhance F formation and eliminate polymerization. Second, with sufficient O_2 present, $O_2(O)$ reactions with the film being etched can inhibit etching as shown by the hysteresis curve (Figure 12) observed in the CF_4/O_2 system (*42*). Similar results were obtained with other gases (*48,50*).

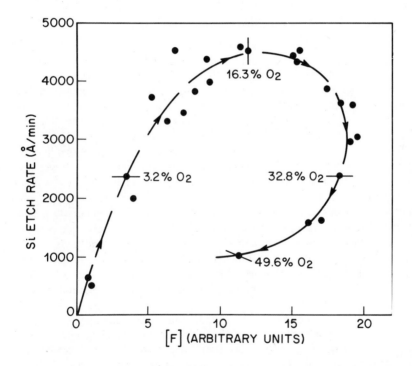

Figure 12.   Silicon etch rate versus F-atom concentration.   Arrows indicate increasing O_2 concentration in the feed.   (Reproduced with permission from Ref. 42.)

Plasmas that produce chlorine and bromine atoms have been shown to be excellent for Si etching, and can display a high degree of anisotropy control. The most commonly used gases have been Cl_2, CCl_4, CF_2Cl_2, CF_3Cl, Br_2 and CF_3Br (51) along with mixtures such as Cl_2/C_2F_6, Cl_2/CCl_4, and C_2F_6/CF_3Cl. Of these, the most useful systems appear to be Cl_2 (52), Cl_2/C_2F_6 and CF_3Cl/C_2F_6 (53,54) where high etch rates (500 − 6300) Å/min for undoped and doped poly-Si and selectivity (Si/SiO$_2$ ∼10-50:1) have been observed. The active etchant in these plasmas is likely to be Cl and Br atoms; however, ion bombardment plays a significant role in achieving high etch rates and anisotropy control (see Section 5.5.c.). The high degree of anisotropy that is readily achieved suggests that ion bombardment tends to dominate the etch mechanism by enhancing the reaction with the chemisorbed SiCl$_2$ layer and/or enhancing product volatility. This is borne out by the fact that only a small loading effect is observed in these cases (53), while F-source plasmas (38,39) exhibit strong loading effects. The absence of a significant loading effect and controllable anisotropy makes Cl-source plasmas most attractive for Si etching.

5.4.c.2. Silicon Dioxide and Silicon Nitride.   The etchants described for silicon are suitable for etching oxide and nitride; however, they can be used only in the absence of silicon. Selective etching of these materials represents the single most important application of the chemical models presented in Section 5.4.b. The earliest reported etchant gases for selectively etching SiO_2 and Si_3N_4 in the presence of silicon were CF_4/H_2, CHF_3, C_3F_8, and C_2F_6 (55). Selectivities as high as 15:1 (Si_3N_4/Si) and 10:1 (SiO_2/Si) were achieved. All of these plasmas tend to be fluorine-deficient suggesting that CF_3 was the active etchant, but this has never been confirmed experimentally. It is likely that CF_2 radicals are equally if not more important since carbenes (i.e., CF_2) are notorious for undergoing insertion reactions (56), whereas alkyl radicals (i.e., CF_3) generally participate in abstraction reactions (57). Furthermore, the recombination rates for CF_3 are considerably higher than for CF_2 under plasma conditions, so that CF_2 is likely to be present in much higher concentrations. Certainly both can account for the appearance of products such as SiF_4, CO and COF_x which have been observed by mass spectrometry (58) (see Section 5.6.c.2.) during SiO_2 etching.

In order to achieve selective oxide and nitride etching, additives to F-source plasmas are chosen to make a F-deficient chemical environment. These include H_2, C_2H_4 and CH_4 which are quite efficient F scavengers. The amount of additive necessary remains more an art than a science because oxide and nitride selectivity requires operation in an environment very close to the demarcation between etching and polymerization shown in Figure 10. In fact in some cases (51,59) polymer deposition on Si occurs while active etching of oxide proceeds. Thus, the mechanism for selective oxide and nitride etching may not involve CF_x as active etchants, but as film formers which inhibit Si etching by passivating chemisorbtive sites. In these cases monolayer coverage of the oxide with CF_x radicals might be expected to release F atoms, a proven active etchant, under ion bombardment.

The role of ions in promoting oxide and nitride selectivity has also been established. Discharge conditions (i.e., pressure and voltages) are generally those that favor ion-enhanced reactivity, and ion bombardment is required to initiate etching (59). The absence of a noticeable dependence on substrate temperature (60), minimal loading effects (61), and the sensitivity of etch rate with respect to electrode potential (62) all reinforce the importance of ion bombardment. In the latter study (62), selective etching of SiO_2 over Si by >50:1 was obtained in a CHF_3 plasma using cathode coupling, while selectivities <10:1 were observed at the anode.

Although adjusting plasma conditions remains an art, high selectivity is achievable, and high etch rates for oxide (600-1000 Å/min) and nitride (>1000 Å/min) have been demonstrated (55,60,62,63,64). A note of caution is in order for interpreting the selective etching of nitride films. Generally, stoichiometric Si_3N_4 shows etch characteristics (46) between those of Si and SiO_2; however, reported selective nitride etching generally exhibits much higher etch rates than SiO_2 under the same conditions. Plasma-deposited nitride, a Si-N-H compound, is far more reactive (46). Thus, some reports (65) of selective nitride etching in F-rich plasmas should be interpreted with caution when a characterization of the film is not given.

5.4.d. III-V Materials. Group III-V semiconductor compounds such as GaAs, GaP and InP form the basis for many new electronics applications, particularly optoelectronic devices. The development of plasma pattern delineation methods is an area of very active research because of the parallelism with the more highly developed silicon technology; however, these systems are somewhat more complicated. Since most of the applications involve binary, ternary and even quaternary alloys, suitable etchants must be capable of selective etching when stoichiometric changes are made to obtain a specific device characteristic. At present only the binary compounds have been studied, and product volatility considerations are the most significant driving force behind much of this research. The III-V compounds are particularly difficult since the group V elements form volatile halides, whereas group III halides, particularly the fluorides, tend to be involatile. As a result, F-source plasmas, which have been the mainstay of silicon technology, are not practical for III-V etching. Therefore, most studies have utilized chlorine-containing plasmas with elevated substrate temperatures to take advantage of the volatility (albeit limited) of the group III chlorides. The importance of the volatility concept in III-V etching relative to the reaction concepts which dominate Si etching has been demonstrated in the etching of InP and GaAs in Cl_2 plasmas (66). Activation energies obtained by measuring etch rates as a function substrate temperature were in excellent agreement with the heats of vaporization of $InCl_3$ and $GaCl_3$ suggesting that vaporization of the group III halide was rate limiting. In fact, the etch rate of InP could be explained entirely on the evaporation rate of $InCl_3$ (66). Although this was not possible for GaAs etching, and the agreement between activation energy and heat of vaporization of $GaCl_3$ was probably fortuitous, the study does highlight the importance of product volatility in etching binary and more complex alloys.

The active etchant for III-V materials is undoubtedly atomic chlorine, and a number of investigations (67-69) into appropriate sources have been conducted. An extensive study (67) has been conducted for the etching of GaAs and its oxide. With the exception of CCl_2F_2, fluoro- and chlorofluorocarbon plasmas were ineffective in etching GaAs, most likely due to reduced volatility caused by fluorine passivation. Chlorocarbon plasmas such as C_2Cl_4, CBr_2Cl_2 and $CHCl_3$ were also ineffective because of polymer deposition; however, CCl_4, CCl_2F_2, PCl_3 and HCl plasmas etched both GaAs and its oxide favoring GaAs in a ratio of 400:1. Finally, Cl_2 and $COCl_2$ plasmas were found to etch GaAs exclusively with very high etch rates (5 μm/min). Here the results can be interpreted in terms of chemical selectivities, since gas pressures and plasma frequency were high enough to minimize ion bombardment effects.

Other studies have been conducted under conditions of low pressure favorable to ion enhanced etching. Anisotropic (68,69) and directional (oblique angles of incidence of ions) (70,71) etch profiles were observed in accord with the concepts of Section 5.5. The role of additives has not been studied to any great extent; however, CCl_2F_2/O_2 plasmas (67,70) enhanced GaAs etching in general and over native oxide in particular while H_2 additions (67) enhanced oxide etching. The origin of these effects is thought to be CCl_2 (67) and its preferential reactivity with oxide, but at this time the interpretation is speculative.

At present Cl_2 appears to be most suitable for etching at moderate pressures since etch rate and anisotropy can be controlled by plasma frequency (72); however, more recent studies (66b,c) suggest that Br_2 plasmas may be superior in this aspect. One interesting feature of using Cl_2 and Br_2 is the observation of anisotropic etching under nominally isotropic plasma conditions (high pressure, high frequency, low power). Unlike the mechanisms for anisotropy discussed later in Section 5.5, anisotropy in this instance was attributable to chemical attack along specific crystallographic planes (66c) containing a higher lattice content of the more volatile group V element. Very smooth vertical walls were observed; however, it was clear from examining the overlaying mask that etching was chemical (isotropic).

5.4.e. Metals. Because of the high reactivity of most metals with oxygen and water vapor, plasma etching of metals often requires more attention to reactor design and process details than is necessary with other materials. Unless the metal forms a volatile oxychloride or an unstable oxide, water vapor and oxygen must be excluded from or scavenged in the plasma reactor. In addition, the metal-oxygen bond may be extremely strong, thereby requiring ion bombardment to assist native oxide removal.

As a result of ion bombardment effects and of their ability to reduce native oxides chemically, chlorocarbon or fluorocarbon gases, rather than pure halogens, are typically used to etch metal films (Table IV).

Table IV. Typical Plasma Etchants Used for Metal Films

Metal	Etchants	Reference
Al	BCl_3, CCl_4, and mixtures with Cl_2, $SiCl_4$	73-75
Au	$C_2Cl_2F_4$, $CClF_3$	73,76,77
Cr	Cl_2, CCl_4, each with O_2	73,78
Ti	CF_4, $CClF_3$, $CBrF_3$	77,79
W	CF_4, SF_6 and mixtures with O_2	72,80

Unfortunately, halocarbon vapors are particularly susceptible to polymerization, causing residue formation that can interfere with etch processes (*74*).

5.4.e.1. Aluminum. Since aluminum is widely used as an interconnect layer for integrated circuits, its plasma etch characteristics have been extensively studied. An initiation period or a lag-time exists at the start of aluminum etching. This "slow etching" period results from the fact that two processes must occur prior to the etching of an aluminum film (*81*): scavenging or removal of oxygen and water vapor present in the reactor, and etching of the thin (~30 Å) native aluminum oxide layer always present on the aluminum surface. The former concern can be minimized by using a "load lock" so that the chamber is not exposed to air (water vapor) between etch runs (*82*), or by using an etch gas (BCl_3, or $SiCl_4$) which effectively scavenges water and oxygen. Native aluminum oxide can be etched by enhancing ion bombardment of the surface and by supplying chemical species capable of reacting directly with the oxide, i.e., CCl_x, BCl_x, or $SiCl_x$.

Since AlF_3 is not volatile under normal plasma etching conditions, chlorine-containing gases have been principally used to etch aluminum. The etch product, $AlCl_3$, is hygroscopic, and has a relatively low volatility at room temperature. Etching with brominated gases has been reported, albeit with poor reproducibility and control (*83*). Chlorine gas has not been successful in reproducibly etching aluminum, primarily because of its inability to etch the native oxide. With BCl_3, oxygen and water vapor scavenging is more effective than with CCl_4 (*81*). Results with $SiCl_4$ as an etchant gas suggest that it behaves similar to BCl_3 (*75*) in scavenging ability.

After removal of the native aluminum oxide layer, molecular chlorine (Cl_2) can etch pure, clean aluminum without a plasma (*84,52*). Indeed, Cl_2 may be the primary etchant species for aluminum in a glow discharge, perhaps due to a higher concentration of Cl_2 than chlorine atoms in the discharge (*74*).

In order to prevent aluminum from spiking through shallow junctions, 1-2% silicon is often added to the film. Since $SiCl_4$ is volatile at room temperature, aluminum-silicon films can be readily etched in chlorine-containing gases.

Copper additions to aluminum films enhance electromigration resistance. Unfortunately, copper does not form volatile chlorides or other halides, thereby making removal during aluminum plasma etching difficult. Two methods can be used to promote copper chloride desorption: increase the substrate temperature (consistent with the resist material being used), or enhance the ion bombardment occurring so that significant sputtering and/or surface heating is attained.

After plasma etching is completed, aluminum films often corrode upon exposure to atmospheric conditions. The corrosion is a result of the hydrolysis of chlorine or chlorine-containing residues (mostly $AlCl_3$) remaining on the film sidewalls, on the substrate, or in the photoresist. Since the passivating native oxide film normally present on the aluminum surface has been removed during etching, chlorine species are left in contact with aluminum, ultimately causing corrosion. Further, carbon contamination and radiation damage caused by particle bombardment may enhance corrosion susceptibility (*85*). A water rinse or an oxygen plasma treatment after etching lowers the amount of chlorine left on the etched surfaces, but is not adequate to preclude corrosion. Low temperature thermal oxidations in dry oxygen appear effective in restoring a passivating native aluminum oxide film (*85*). Another method of preventing post-etch corrosion is to expose the aluminum film to a fluorocarbon plasma (*86*). This treatment converts the chloride residues into nonhygroscopic fluorides. Subsequently, a nitric acid rinse can be used to remove the fluoride layer and to regrow the protective oxide.

Chlorine-based plasma etching of aluminum films causes serious degradation of photoresist materials. To some extent, these effects are a result of the etch product, $AlCl_3$. Aluminum trichloride is a Lewis acid used extensively as a Friedel-Crafts catalyst. Therefore, it is hardly surprising that this material reacts with and severely degrades photoresists (*74*).

5.4.e.2. Other Metals. Numerous other metal films have been etched in a glow discharge. The following paragraphs present specific information on the etching of the metal films indicated in Table IV.

Gold can be etched effectively with $C_2Cl_2F_4$ (76) or with $CClF_3$ (77), while CF_4/O_2 etching causes staining. The observed staining is believed to be gold oxides, whose formation is enhanced by the presence of atomic fluorine (77).

Chromium is etched readily in plasmas containing chlorine and oxygen (78) because of the high volatility of the oxychloride ($Cr_2O_2Cl_3$). Indeed, the high boiling point of $CrCl_2$ (1300°C) results in significantly reduced etch rates of chromium in chlorine plasmas without oxygen.

Titanium can be etched in fluorine-, chlorine-, or bromine-containing gases, because all the halides are volatile. Chlorides and bromides have been studied to a great extent since they result in high selectivity over silicon- containing films, and do not promote staining on gold (77).

Tungsten films can be etched easily in fluorine-containing plasma (73,80). If carbon or sulfur is present in the gas atmosphere, oxygen is typically added to prevent polymer and residue formation and to increase the concentration of fluorine atoms.

5.4.f. Organic Films. Organic films are present during the plasma processing of all materials discussed in preceding sections since polymeric resist masks are the primary method of pattern transfer. An ideal mask should be highly resistant to the reactive species, ion bombardment, and UV radiation produced in the glow discharge. In addition, it should be readily removed once pattern delineation is complete. Unfortunately, very high selectivities of etch film to resist material are not often achieved, particularly in CF_4/O_2 processing of silicon-based materials. As a result, linewidth loss in pattern delineation (*see* Section 5.5) frequently occurs as the thinner mask edges erode during etching. A cursory examination of the literature indicates that for one commonly used group of resist materials (Shipley AZ1350 series) silicon etch rates can be greater than (87), comparable to (34,88), or less than (89) resist etch rates depending on gas composition and discharge parameters. Although selectivities of silicon-to-resist as high as 9:1 have been observed (87), selectivities of 2-5:1 are probably more the rule in CF_4/O_2 plasmas. Other resist materials (90) have demonstrated a high immunity to degradation in a CF_4/O_2 plasma; however, it is not clear whether microwave excitation was responsible for these observations.

As will be shown shortly, CF_4/O_2 plasmas provide a most stringent test for resist durability; however, there are some plasmas in which resist durability is quite high. For example, conditions favorable for the selective etching of SiO_2 and Si_3N_4 are not nearly so conducive to resist degradation as are CF_4/O_2 plasmas. As a result, CF_4/C_2H_4 (91), CF_4/H_2 (92), CHF_3

(34,88,93), and C_2F_6/C_2H_4 (60) plasmas exhibit excellent selectivity of oxide over resist even when ion bombardment is present. Considering that selective oxide etching occurs in a saturate-rich plasma near the borderline of polymer deposition, the selectivity over resist etching is not surprising. Any degradation of polymeric resist material is likely to be compensated by saturate condensation reactions at these sites.

Although a complete discussion of polymer properties leading to highly durable masking materials for plasma etching is beyond the scope of this chapter, one study is most noteworthy. The relative etch rates of a wide variety of polymers in an O_2 plasma exhibited a high correlation between structural properties of the polymer and their stability when exposed to the plasma (94). More importantly, the results indicated a synergistic degradation involving atomic oxygen from the plasma and halogen present in either the polymer or the plasma (i.e., from CF_4). Based on these studies, it is quite clear why CF_4/O_2 discharges produce high mask erosion rates. Atomic fluorine abstracts hydrogen from the mask, producing sites that react more readily with molecular oxygen. The net result is an increase in polymer etch rates over those due to atomic oxygen alone. Recent EPR (see Section. 5.6.d.3.) studies (95) of photoresist stripping downstream from a microwave discharge have confirmed the role of O and F atoms in this mechanism. On the other hand, oxygen-free plasmas tend to result in polymer stabilization. Halogen abstraction of polymer hydrogens followed by reaction with halogen or halocarbon radicals leads to halocarbon groups in the polymer which make the mask more resistant to plasma degradation.

Although the etch characteristics of other materials can be understood primarily in terms of reactivity with halogen atoms, etching of organic films is dominated by reaction with atomic and molecular oxygen. Oxygen plasmas provide a highly selective medium for removal of organic materials and have been used extensively for stripping photoresists (96) and removing epoxy smears from other electronic components (97). More recently, the use of oxygen plasmas in delineating the original mask pattern itself has become a very exciting field. The simple concepts of a plasma-developed resist process are depicted in Figure 13 (98). A moderately volatile monomer (m) in a film of host polymer (P) is exposed to ionizing or UV radiation to lock the monomer to the host polymer (P-m linkages). Following a baking cycle to remove free monomer in the unexposed regions, an appropriate plasma is used to selectively etch either the parent polymer (P) or the copolymer (P-m), leaving a negative or positive resist mask for plasma processing the underlying film. Usually the copolymer is more resistant to an O_2 plasma and the negative tone results (98). Other chapters deal more thoroughly with this emerging technology; however, gas mixtures and plasma conditions for achieving selective polymer etching will be necessary to exploit it fully.

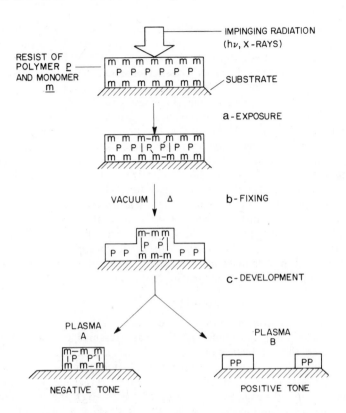

Figure 13. Schematic of plasma-developed resist film composed of polymer host (P) and volatile monomer (m). Processing steps are (a) exposure which locks monomer in place, (b) fixing which removes unlocked monomer, (c) plasma development. (Reproduced with permission from Ref. 98a.)

Plasma conditions that favor ion bombardment can be used to enhance resist etch rates and to obtain anisotropic resist profiles. Reactive sites produced by bombardment permit more rapid attack by atomic and molecular oxygen in the plasma. Tri-level processing (*see* Chapter 6), in which an oxide mask is first used to pattern a polymer layer before transferring the pattern to an underlying semiconductor film, relies heavily on such anisotropic resist development. Ultraviolet light which is present in all glow discharges also enhances the purely chemical etchant activity and is particularly important for organic film materials. The photochemistry of organic molecules (99) is a well established field. Although polymeric films are a more complex photochemical system, many of the chromophoric groups in the polymer react similarly under exposure to UV radiation. The most serious degradation results when scission of backbone or side-chain (near the backbone) bonds of the polymer occurs. Active sites become available for reaction with atomic and molecular species in the plasma. In order to stabilize these bonds, polymer scientists specifically design side groups far removed from the backbone to bear the brunt of photochemical attack.

Unlike silicon-based materials where selective reactants are of ultimate importance, and III-V and metallic materials where product volatility dominates etching considerations, selective etching of organic films is driven by incorporating the desired reactivity (or lack of it) into the film itself. In device fabrication all types of materials are present simultaneously and the process engineer must be aware of the important aspects of the chemistry of each material in addition to the gas phase reactions that produce chemically active species. It is hoped that the discussions presented here provide a basis for approaching such a complex chemical system and for critically evaluating studies which appear in the literature.

5.5 Profile and Dimensional Control

5.5.a. General Considerations. The goal of any pattern etching process is to transfer an exact replica of the mask features to the underlying film. However, this establishes only a two-dimensional criterion for the quality of the replication. With the cross-sectional etched features shown in Figure 14, the quality criterion can be set down in rather simple terms. In each case the initial projected patterning of spaces (S_o) and linewidths (W_o) is identical, yet the final dimensions (S_F, W_F) show a potential for considerable variation. It is clear that totally anisotropic etching will satisfy the replication criterion $(S_F=S_o, W_F=W_o)$ exactly, while isotropic etching will introduce linewidth loss $(W_F = W_o - 2d_V, d_V=d_H)$ and a concomitant increase in feature separation $(S_F = S_o + 2d_V)$. Thus, dimensional control will depend on a knowledge of the degree of anisotropy that can be achieved or at least measured for a given etching medium. The most useful measure is the ratio of lateral undercutting (d_H) to vertical etching (d_V) which exhibits an inverse relationship to the quality of dimensional replication. In anisotropic etching, $d_H/d_V = 0$ and high quality dimensional transfer is obtained. Isotropic etching, where $d_H/d_V = 1$, gives a low quality transfer. Of course, in isotropic etching the final feature dimensions can be controlled to yield desirable pattern delineation (W_F, S_F) by increasing the mask width by $2d_V$ prior to etching. Such an approach is useful in low density patterning; however, VLSI applications $(W_F, S_F \leqslant d_V)$ render this approach impossible. Thus, the development of anistropic etching techniques is crucial for fabricating high density devices.

Although the ability to control the two-dimensional pattern is important, an additional criterion for judging the quality of etching is dimensional variation in the vertical direction. Unlike the transfer of dimensional information from mask to film, no single ideal can be established for edge profiles since the desired profile depends largely on specific application in the final device. If a conductor line is being defined for subsequent metallization, a steep walled profile is desirable to maximize the cross-sectional conductor area and minimize coverage of the step. Here, anisotropic etching satisfies both criteria, and very fine-lined features can be etched with dimensions approaching the limits of resist lithography ($S_o = W_o \sim 0.1 \ \mu m$ (100)). On the other hand, if uniform step coverage is desired, a tapered profile like that in Figure 14 is essential since highly anisotropic profiles

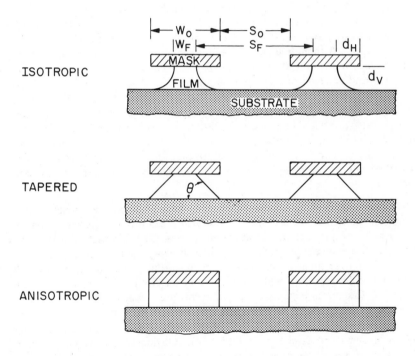

Figure 14. Etch profiles for isotropic, tapered, and anisotropic etching of a film. S_o, W_o and S_F, W_F represent mask dimensions before etching and feature dimensions after etching, respectively. The degree of undercutting (d_H) and wall taper (θ) are indicated for etching to a depth (d_v) that exposes just the initial mask dimensions in the substrate. (Reproduced with permission from Ref. 11.)

tend to cause shadowing during metal deposition leading to a thinning of the conductor at the etched steps. To obtain such a profile and control the degree of taper (θ) one must be able to introduce isotropy into an anisotropic etch and adjust plasma conditions to achieve any desired degree of anisotropy ($= 1 - d_H/d_V$). Of course, this may mean sacrificing control in pattern delineation or adjusting mask dimensions to compensate for the isotropic component as noted earlier.

A major factor in pattern delineation and profile control in an etch process is the mask material and the rate of erosion when exposed to the plasma. The masks shown in Figure 14 represent an ideal case which is achievable using multi-level processing (Chapter 6). In many applications single level resists are used, and because of diffraction limitations they tend to be notably thinner at the edges of a feature. As a result, considerable linewidth loss may occur in an anisotropic etching medium due to erosion of the thinner regions of the mask. However, such erosion can lead to tapered profiles which may be desirable. In fact, it is possible to control taper of the mask near a step and transfer a tapered profile to the film during etching by mask erosion. Although such a procedure is often attempted in low density applications, it requires stringent control of all plasma parameters (temperature, ion and electron energy distributions, etc.). Therefore, control of this process over the surface of a single wafer and especially over a batch of wafers is difficult. As a result, it is more prudent to develop etch techniques that permit variability in degree of anisotropy which can be tailored to produce the desired edge profile. In the subsequent parts of this section, conditions and mechanisms leading to isotropic and anisotropic etching are described and an example is given where the degree of anisotropy has been controlled.

5.5.b. Isotropic Etching. Isotropic etching in a plasma arises strictly from chemical reaction of an etchant species with the film. The plasma plays no role other than production of the active species. If the etched film is structurally and chemically isotropic as well, then every point on an exposed surface will etch at an identical rate. The resulting profile for a just-etched feature (i.e., Figure 14) will exhibit a quarter-circle cross-section with an origin at the mask edge. Here, the term "just-etched" refers to etching which exposes a line of width S_o in the underlying substrate. Continued etching (overetching) produces concentric circular arcs (*101*) of increasing radius of curvature as shown in Figure 15. Obviously, an overetch greater than 300% (i.e., x/h > 4) will appear as an anisotropic etch when the mask is removed prior to microscopic examination of the profile. Similarly, many masks exhibit erosion during isotropic etching which gradually enlarges the mask opening. With significant erosion, the overetched feature may appear anisotropic even without removing the mask. Therefore, profiles should always be interpreted by comparing the final feature dimensions with those initially on the mask.

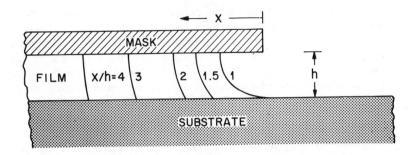

Figure 15.   Isotropic etching a film (thickness = h) showing undercutting (x) of the mask.   Overetching from 0% (x/h = 1) to 300% (x/h = 4) results in profiles which appear more anisotropic.

Although anisotropic etch processes are often desirable, some applications for isotropic etching exist. For example, either cleaning the side walls following a sputter etch (physical ablation of material) or producing a slight undercut is best accomplished with an isotropic etch. Furthermore, it is absolutely essential when material underneath a nonetchable overlayer must be removed (*39*). The best assurance for achieving total isotropy in an etch is to generate the reactive species upstream from the etch zone (*102*).

5.5.c. Anisotropic Etching — The Role of Ions.   Some of the earliest examples of anisotropic etching were obtained using sputter etching techniques (*11*). Here, energetic rare gas ions impinge on a film, and material is physically ablated as a result of momentum transfer. Because of the highly directional nature of the accelerated ions, a high degree of anisotropy can be obtained. Unfortunately, there are several problems with the sputtering approach to anisotropy. First, ablated material from chamber walls and from deep within an etched feature tends to redeposit. Second, sputtering is only weakly selective. After a certain time, the redeposited material becomes a mixture of film, mask, and cathode materials, and subsequent sputtering exhibits no preference for mask, film, or substrate. Finally, etch rates tend to be rather low -- on the order of a few hundred Å/min (*11*) for most materials.

When ion bombardment occurs in the presence of chemically reactive species (plasma etching), the above problems can be alleviated. It would be desirable to establish specific discharge parameters and gas compositions for obtaining any desired profile for any given material. Unfortunately, sufficient fundamental information to define such processes for the wide variety of reactor geometries and rf coupling configurations is not available. The best that can be done is to give the reader a phenomenological description of, and guidelines for anisotropic etching in a system in which a combination of physical and chemical material removal processes are active.

In a previous section (5.4.b.) the synergistic interaction of XeF$_2$ chemical beams and ion beams in Si and SiO$_2$ etching was noted. These effects appear to be very general phenomena in plasma etching. In addition to enhancing etch rates, this synergism is also important in introducing considerable anisotropy into the isotropic, chemical component of the etch process. A very simple, general picture to explain these results is indicated by the two mechanisms (*53,103,104*) in Figure 16.

In the ion induced damage mechanism energetic ions break crystal bonds on the film surface thereby making the film more accessible and more reactive to the active chemical etchant. However, the side walls remain relatively unperturbed, and etching proceeds at the nominal chemical etch rate. Consequently, material removal proceeds far more rapidly in the ion flux direction, resulting in anisotropy. In actuality, the surfaces exposed to the plasma are likely to be composed of a chemisorbed coating of etchant

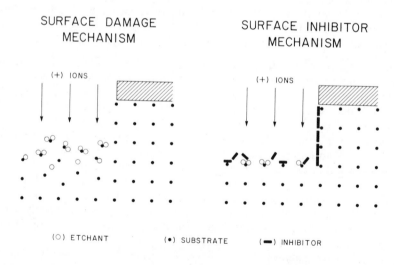

Figure 16. Surface damage and surface inhibitor mechanisms for ion-assisted anisotropic etching. (Reproduced with permission from Ref. 103.)

radicals as well as unsaturate species (*see* Section 5.4.b.) which inhibit the etch rate. Ion bombardment can stimulate the desorption of these species so that surface coverage caused by inhibitors is smallest in regions of high ion flux (Figure 16). Again, increased anisotropy is the net result. The damage induced mechanism requires considerably more energy than the surface inhibitor mechanism for anisotropy, and both are likely to play a role in plasma etching processes. The degree of anisotropy will depend on ion flux and ion energies in the plasma.

As noted earlier, ion flux and ion energy depend very strongly on reactor design and gas composition, and are therefore virtually impossible to translate directly from one reactor to another. However, the important parameters can be qualified to some extent. Ion bombardment is enhanced by decreasing pressures in a high frequency ($\geqslant$ 5 MHz) plasma or by decreasing the frequency of the plasma discharge. In the former case, the ion energies tend to be low since they are exposed to the accelerating voltage of the plasma for only a short period of time. Low pressures are required to reduce the extent of moderating or reactive collisions of the ions with neutral species in the sheath. Surface damage and surface inhibitor induced anisotropy are possible; however, as pressure increases and ion energies moderate because of collisional processes, the inhibitor mechanism becomes more favorable. Under these conditions, it should be possible to achieve anisotropy by adding film forming precursors to the plasma. At low frequencies (1kHz-1MHz), ions are strongly accelerated because of the longer duration of the accelerating potential and both anisotropy mechanisms are operable to much higher gas pressures ($\sim$0.3 Torr).

There is an increasing body of experimental data to support the above contentions (*103*). In high pressure (0.3 Torr), high frequency (13.6 MHz) Cl_2 plasmas, isotropic etching of poly-Si is observed; however, addition of film-forming precursors (C_2F_6) leads to anisotropic profiles (*53*). Similar results were observed in CF_3Cl/C_2F_6 plasmas (*105*). Anisotropic etching of poly-Si in Cl_2 plasmas was observed (*106*) at similar pressures at low frequencies (100 kHz). Of course, in the low pressure regime, anisotropic etching of Si in Cl_2 plasmas (*107*) is obtained routinely.

A most interesting demonstration of these concepts in profile control are the recent results obtained with ClF_3/Cl_2 mixtures (*39*). At a fixed pressure of 0.02 Torr and 100 W of 13.6 MHz rf power, mixtures of varying composition were used to generate a continuous spectrum of profiles with varying anisotropy. Figure 17 (*39*) shows the degree of isotropy (d_H/d_V as defined in Figure 14) as the ClF_3 concentration is varied. In chlorine-rich plasmas (% ClF_3 <10) etching is highly anisotropic ($d_H/d_V \rightarrow 0$), while in fluorine rich plasmas d_H/d_V approaches one as expected for the isotropic nature of F atoms as an etchant. Thus, in this system, as well as others (*53,54*), arbitrary profile control is possible while maintaining high etch selectivity.

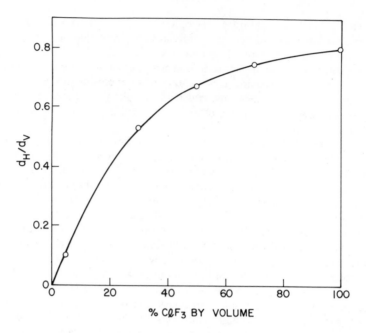

Figure 17.   Degree of isotropy observed in silicon etching in Cl_2/ClF_3 plasma as a function of ClF_3 content of the feed.   Conditions are 5 sccm total flow, 0.02 Torr pressure, 100 W r.f with 30 V cathode bias at a frequency of 13.6 MHz.   (Reproduced with permission from Ref. 39.)

Although increasing rf power has been the traditional approach to increasing ion energies and anisotropy, the more subtle parameters of plasma frequency, voltage and composition lead to very dramatic effects. The guidelines and mechanistic views presented here should be used only as a starting point for the reader to pursue specific conditions for achieving specific results.   More detailed information can be gathered from the citations given elsewhere (*103,108*).

5.6 Process Monitoring (Diagnostics)

5.6.a. Need for Diagnostics.   As can be seen from previous discussions, dry etching using plasma processing involves a wide variety of feed gas compositions, substrates and physical variables such as rf power, gas flow rates, pressure and temperature.   Although the "black-box" approach of changing a variable and observing effects can be taken in optimizing a single process, it is far better and, in the long run, far more expedient to identify and understand individual processes that occur within the overall process.

From the point of view of process development, a detailed understanding of one physico-chemical system is imperative for insight into the design of processes for etching new substrate materials. To meet these needs, diagnostic probes are essential for identifying and monitoring the concentration of active etchants in the sea of potentially reactive species formed in an rf discharge. The ability to identify and monitor etch reactants and products allows the process engineer to unravel the kinetics and detailed mechanism of any etch process. In addition, the important plasma variables for optimizing the concentration of an active etchant, achieving etch selectivity of one material over another, and achieving adequate linewidth and profile control can then be identified. Equally important, the application of diagnostic probes should permit one to identify and measure process variables in such a way that the results obtained in one plasma reactor can be directly translated and applied to any other reactor.

The need for diagnostic tools for production reactors is paramount to ensure high yields of reliable devices. Extremely tight control of all process parameters must be maintained to ensure wafer-to-wafer and run-to-run reproducibility. In a typical production environment some of these parameters can be controlled, albeit with great difficulty, while others cannot be controlled at all. For example, reactor wall conditions which contribute to the heterogeneous destruction of active etchants are truly a variable from run-to-run. Similarly, out gassing, residual leaks and backstreaming from vacuum pumps can alter the process chemistry enough that the calibrated etch-time approach to reproducibility can be seriously compromised. Thus, a process monitor for determining the end point of an etch cycle is extremely valuable for reducing the need for overetching, and increasing throughput and run-to-run reproducibility. Although the actual diagnostic tools for end-point detection are essentially the same as those employed for process development, considerable simplifications can be introduced for process control since one is generally seeking a yes/no type of response. Several types have been developed and generally involve monitoring an etchant, product, or the film being etched. A recent review and comparison (*109*) of the various methods of end point detection is available.

In this section, the underlying principles behind some of the most widely used plasma diagnostic methods are presented along with examples relevant to the basic understanding of plasma etching and process end point detection. These include direct film measurements using interferometry and reflectivity, gas phase monitoring of etchants and etch products using optical emission and mass spectrometry, and measurements of the electrical properties of plasmas using Langmuir probes. In addition, the more specialized and less frequently used techniques of laser induced fluorescence and electron paramagnetic resonance spectroscopy are briefly described. Both methods offer unique abilities for monitoring short-lived species in low concentrations and should become valuable tools for increasing our understanding of plasma processes. It is hoped that the discussion will not only give the reader an understanding of each method, but also provide some insight into their advantages and disadvantages so that a critical evaluation can be made for choosing one diagnostic method over another.

5.6.b. Direct Film Measurements.

5.6.b.1. Stylus Surface Profiling. The simplest diagnostic is the stylus surface profile measurement which provides an absolute determination of etch rate for any given set of process parameters. The procedures are illustrated in Figure 18a. A suitably masked substrate is exposed to the etching plasma for a measured time. Following etching, the mask is removed and the surface profile of the etched substrate is measured by tracking a small stylus over an etched feature (Figure 18b). A sensitive strain gauge senses the change in strain and an amplified signal corresponding to the change (H) is recorded. By comparing the etched sample to standard profiles the etch depth can be determined with an accuracy consistent with any roughness or texture introduced during etching. Most stylus profiling instruments are capable of measuring etch depths from 50Å to 100μm; however, considering texture introduced during etching, a few hundred Å represents a more reasonable lower limit. Knowing the duration of the etch cycle, one can readily compute the absolute etch rate in Å/min or μm/min. Although a simple measurement, stylus profiling is generally very reliable and not prone

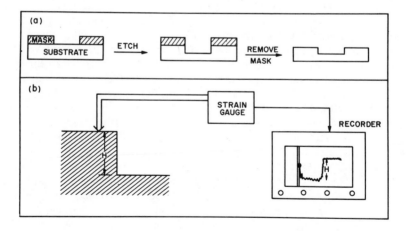

Figure 18. Experimental procedure (a) and stylus thickness measurement (b) for determining etch depth.

to serious error when more than 1000Å of material has been removed. The most serious disadvantage is that it is not an *in situ* technique and etch rate determinations for a variety of process parameters require extensive amounts of time and samples.

5.6.b.2. Optical Reflection and Interferometry. Plasma etch monitoring using optical reflection (*110*) and laser interferometry (*110,111*) has proven to be a very simple and useful *in situ* film thickness monitor and end point detector. The experimental diagram in Figure 19 shows a simple arrangement for measuring the intensity of light reflected from a film whose etch characteristics are being studied. Considering the three layer system of Figure 19 with refractive indices n_o, n_1 and n_2, the theoretical reflectivity of such a system has two useful limiting cases. If the film being etched is metallic the complex index of refraction ($n_1 = n_1 + i\kappa$, κ = extinction coefficient) must be used and the reflectivity for normal incidence is given by

$$R = \frac{(n_o - n_1)^2 + \kappa^2}{(n_o + n_1)^2 + \kappa^2} \tag{20}$$

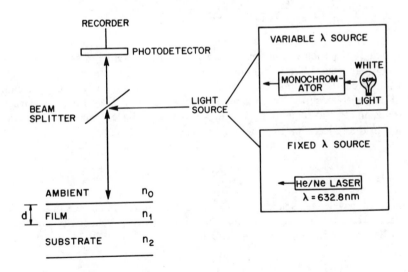

Figure 19. Configuration for experimental reflectivity measurements on a film using a fixed or variable wavelength light source. The film thickness (d) is indicated as well as the refractive indices of the plasma (n_o), film (n_1), and substrate (n_2).

This expression is independent of the film thickness. Thus, when one considers reflective monitoring of metal film etching, only at the interface between film and substrate will a change in reflectivity be observed due to the change in refractive index. Although this is extremely useful for end point detection one still must apply films of known thickness for cases in which etch rate information is desired.

In the ideal case one should observe an abrupt change in reflectivity as the interface between two metals clears during etching. However, two things can obscure the end point. First, for highly reflective metal films the reflectivity is dominated by κ in Equation 20 so that $R \approx 1$ for many metals, and the interface will correspond to only a small change in reflectivity. Secondly, any texturing or tarnishing of the metal during etching can cause scattering or abnormal reflectivity changes which might obscure the end point. Despite these potential drawbacks, the technique has been successfully demonstrated (110) for a gold ($R \approx 0.9$)/tantalum ($R \approx 0.5$) metallization.

Reflectivity measurements as an *in situ* etch monitor is an extremely powerful tool when the film under study is a non-absorbing dielectric such as SiO_2. In this case the reflectivity is given by (112)

$$R = \frac{n_1^2(n_o-n_2)^2 + (n_o n_2-n_1^2)^2 + (n_o^2-n_1^2)(n_1^2-n_2^2)\cos(4\pi n_1 d/\lambda)}{n_1^2(n_o+n_2)^2 + (n_o n_2+n_1^2)^2 + (n_o^2-n_1^2)(n_1^2-n_2^2)\cos(4\pi n_1 d/\lambda)} \quad (21)$$

which is a periodic function of $2\pi n_1 d/\lambda$, and therefore d and $1/\lambda$. As a result, sinusoidal-like variations in the reflectivity are observed when a continuous change in film thickness (i.e., during etching) or source wavelength (Figure 19) occurs. Varying the source wavelength is useful for accurately determining the remaining film thickness since adjacent maxima (at λ_1 and λ_2) in the reflectivity give the film thickness as

$$d = \lambda_1\lambda_2/2n_1(\lambda_1-\lambda_2) \quad (22)$$

A fixed wavelength source is useful for determining the change in film thickness. Here the adjacent maxima (or minima) in the reflectivity are related to the change in film thickness (Δd) by

$$\Delta d = \lambda/2n_1 \quad (23)$$

The varying reflectivity is associated with the constructive and destructive optical interference produced by reflections from the front and rear (interface) surfaces of the film; thus, the general term interferometry is applied to such measurements.

The ability to monitor the change in film thickness during etching is an extremely powerful technique for rapidly ascertaining the absolute etch rate of a dielectric film as plasma parameters are varied. For example, Figure 20 shows the reflectivity (at $\lambda = 632.8nm$) of a polysilicon film on an SiO_2 substrate as a function of time during etching in a CF_4 plasma (110). The etch time between adjacent maxima is seen to be ≈ 78 sec. With $n_1 = 3.85$ for silicon and Equation 23, an etch rate of 649Å/min (110) can be determined on the basis of the separation of the two adjacent minima indicated. The method is useful for absorbing films such as Si (as in Figure 20) provided they are reasonably thin and the complex refractive index is used; however, for thicker films it is probably better to use a non-absorbed wavelength. For source wavelengths longer than $1\mu m$ silicon is relatively

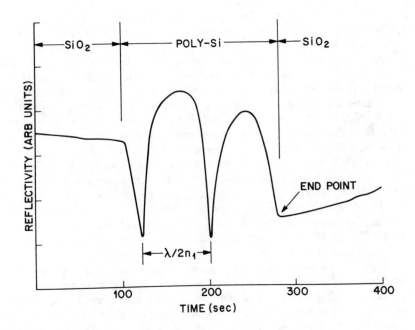

Figure 20.   Reflectivity variations during the etching of a poly-silicon film on an SiO_2 substrate. (Reproduced with permission from Ref. 110.)

nonabsorbing and Equation 23 can be applied directly. It can be seen that by using a wavelength of $1\mu m$ the change in film thickness corresponding to one interference fringe will be 1300Å and at $\lambda=5\mu m$, $\Delta d = 6500\text{Å}$. Thus, longer wavelengths tend to reduce the accuracy of etch rate determinations unless very thick films are used. For other film materials the source wavelength chosen should be one that is not absorbed and for which the refractive index is known.

As an end point detector, interferometry provides very distinct transitions. At the interface between two dielectrics (see Figure 20) the end point is indicated by a change in slope and frequency in the reflectance variations caused by the different refractive indices and etch rates of the two materials. Metal/dielectric interfaces are marked by a variation between an approximately constant reflectivity and an oscillating one. Laser interferometry can monitor regions as small as $0.5mm^2$ (109) on a single wafer or much larger areas with beam expansion. The major problem is that only single wafer monitoring is possible, and erroneous results in end point detection might occur in a production reactor containing many wafers.

5.6.c. Gas Phase Monitoring.

6.5.c.1. Optical Emission. The characteristic glow associated with an r.f. discharge in plasma processing has been immensely valuable in obtaining an improved understanding of the types of species present and how they are formed. By using a monochromator in an experimental arrangement such as that shown in Figure 21, plasma scientists have been able to disperse the optical emissions responsible for the glow into a wavelength spectrum which is characteristic of species present and of the level of excitation. In practice, optical filters are frequently placed between the emission source and entrance to the monochromator. When emission spectra in the visible region (400-700 nm) are being recorded, an ultraviolet cut-off filter is used. This prevents a UV band at 250 nm, for example, which also appears at 500 nm owing to second-order diffraction from being misinterpreted as a new feature in the visible region. Figure 22 shows the emission spectra from r.f. plasmas of N_2 (a) and a CF_4/O_2 mixture (b). This study (113) established a strong correlation between atomic fluorine emission and the etch rate of silicon (Figure 23), suggesting that atomic fluorine was the active etchant in CF_4/O_2 discharges.

Assignments of the various emission bands in Figure 22 are indicated, and a rudimentary understanding can be gained by comparison with the corresponding energy level diagrams shown in Figure 24. Obviously, molecular species present a more complex spectrum than do atomic species because of vibrations and rotations of the molecule. For example, considering the Second Positive System of N_2 involving the electronic $C^3\Pi_u$ and $B^3\Pi_g$ states, progressions corresponding to the various vibrational levels of

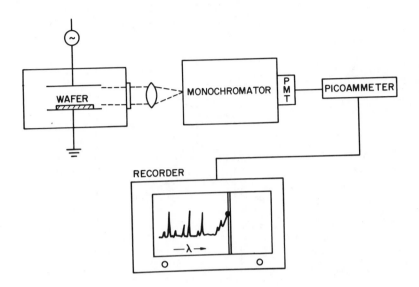

Figure 21. Experimental arrangement for monitoring optical emission from an r.f. plasma. The photomultiplier tube (PMT) and picoammeter detection electronics are frequently replaced with photodiode arrays and photographic film in many spectroscopic studies.

the upper (primed) and lower (double-primed) states are clearly evident, and an even more extensive series of progressions is seen for the First Positive System $(B^3\Pi_g \rightarrow A^3\Sigma_u^+)$. More complex molecular species would, of course, present an even more complex spectrum, and when one considers that mixtures of gases are frequently employed in plasma etching, spectral assignments and identification might appear hopeless. Fortunately, spectroscopists have made a career of studying emission spectra in glow discharges and compendia (114) are available to aid in species identification. Much simpler spectra are observed for atomic species such as F atoms even though a large number of electronic state multiplets (indicated in Figure 24b) are possible. Optical emission is very well suited for species identification since the spectra, complex as they may be, provide a fingerprint of each atom or molecule which can be obtained in a nonintrusive way.

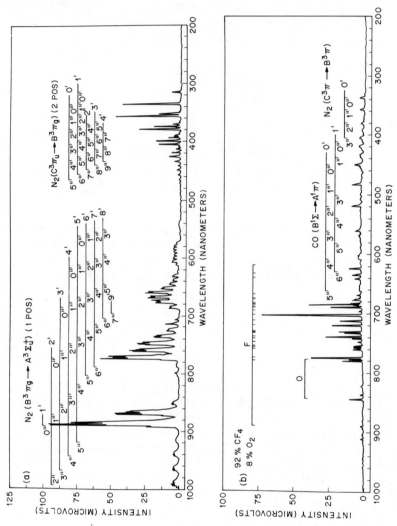

Figure 22. *Emission spectra from r.f. plasmas of (a) N₂ and (b) CF₄/O₂.*
(Reproduced with permission from Ref. 113.)

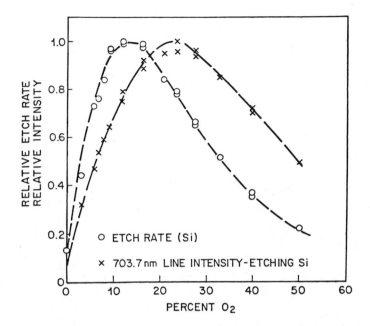

Figure 23. The normalized etch rate of Si *(○) and F-atom emission intensity (×) in a* CF_4/O_2 *plasma as a function of* O_2 *concentration in the feed. (Reproduced with permission from Ref. 42.)*

In addition to identification, optical emission provides an excellent probe into the energetics of the fundamental plasma processes. Even though emission spectra are usually obtained for the UV/visible region (200-800nm) of the electromagnetic spectrum corresponding to changes in energy on the order of 1.5-6.0eV, the energy of the electronic states involved frequently corresponds to much higher levels of excitation. For example, as Figures 22 and 24 show, an observation of the Second Positive System of N_2 or the 703.7nm line (3p, $^2P^o \rightarrow$ 3s, ^2P) of F requires a minimum level of excitation of 11.0eV and 14.5eV, respectively. Similarly, observation of emission from molecular ions such as N_2^+ requires even higher excitation levels.

Since the formation of a given species

$$AX + e \rightarrow A^+ + X + 2e \qquad (24)$$

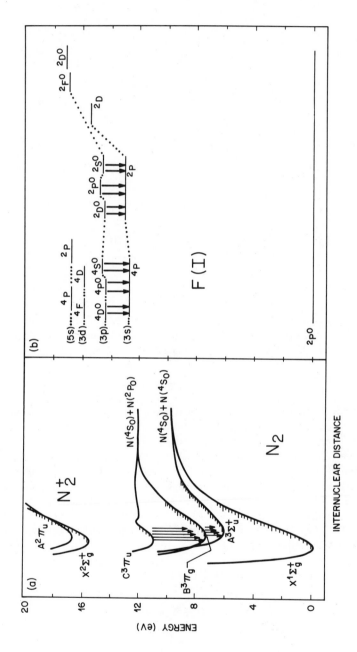

Figure 24.   Energy level diagrams for some of the low lying states of (a) N_2 and (b) F atoms.   Downward arrows indicate some of the transitions giving rise to the features in the emission spectra shown in Figure 22.

and the excitation of emission from it

$$X + e \rightarrow X^* + e \qquad (25)$$

$$X^* \rightarrow X + h\nu \qquad (26)$$

are likely to arise from inelastic electron impact in the plasma, the observation of certain species and excited state emission provides information about electron energy distributions and electron density as well. The dissociation of polyatomic etchants, such as CF_4, utilizes electrons with kinetic energies greater than $\approx 3eV$ (i.e., considerably less than the energy required in Figure 24 to form $N(^4S_o)$ atoms from ground state $(X^1\Sigma_g^+)$ N_2, while excitation of emission utilizes electrons with energies greater than 10eV since emission is often between two excited state levels as just discussed. In other words, the two processes have different energy thresholds and utilize different (yet overlapping) portions of the electron energy distribution. The electron energy distributions shown in Figure 5 for various gas mixtures indicate that dissociations of molecules into neutrals generally involve portions of the distribution function $(\epsilon < 10eV)$, which are relatively insensitive to plasma gas composition for a given frequency, power and bias voltage. On the other hand, the formation of ionic species and excited states (leading to optical emission) involves the high energy tail of the distribution $(\epsilon > 10eV)$ which is extremely sensitive to these variables. As a result, use of the emission intensity as an indicator of the relative concentration of say F atoms can lead to misleading results since the intensity (excitation from the ground state is assumed) will be proportional to the product

$$I \propto kn_e n_F \qquad (27)$$

In other words, emission intensity is a very intimate convolution of the electron energy distribution $(k \propto f(\epsilon))$, electron density (n_e), and the density (n_F) of the species whose concentration is being monitored.

For example, the F-atom emission intensity as a function of feed gas composition in a CF_4/H_2 discharge (*115*) shown in Figure 25 suggests that hydrogen initially destroys F atoms and then, contrary to chemical intuition, creates them. Recent studies (*116,117*) have demonstrated the use of a noble gas actinometer for reducing such potential errors in CF_4/O_2 plasmas. This technique involves adding a small concentration of Ar and monitoring Ar emission $(\lambda = 750nm)$ originating from a state with absolute energy comparable to that involved in F-atom emission. These studies then showed that

Ar emission, which samples the same portion of the electron energy/density distribution, provided an excellent mapping of the excitation efficiency (i.e., kn_e in Equation 27). Therefore, by taking a simple ratio of F/Ar emission intensities, the effects of varying electron distributions can be normalized out of the data. Figure 25 also shows the Ar (λ=750nm) emission data and F/Ar intensity ratios which are in much better agreement with the notion that H_2 is an excellent scavenger of F atoms.

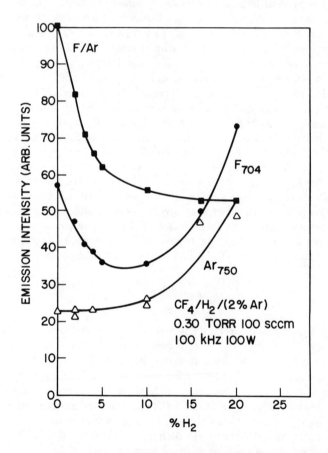

Figure 25. Emission intensity from F atoms (●, $\lambda = 703.7$ nm) and Ar atoms (△, $\lambda = 750$ nm) in a CF_4/H_2 plasma as a function of H_2 concentration. The ratio (■) of F/Ar emissions is also shown. (Reproduced with permission from Ref. 115.)

Similar more quantitative studies (*118*) have been carried out for F atoms in CF_4/O_2 discharges and Br and Cl atoms in Br_2 and Cl_2 discharges, and the accuracy of the actinometer approach over a wide range of plasma composition, frequency and r.f. power is well established. Two basic criteria must be satisfied by a suitable actinometer. First, it must be inert so that its concentration is invariant in the discharge. Second, emission from the actinometer must arise from an energy level in as close proximity as possible to the excited state giving rise to the emission for the species under study. Problems can arise when the mechanism for populating the emissive state does not involve direct excitation from the ground state. If for example, direct dissociative excitation

$$AB + e \rightarrow A + B^* + e \qquad (28)$$
$$B^* \rightarrow B + h\nu$$

or energy transfer

$$A^* + B \rightarrow B^* + A \qquad (29)$$
$$B^* \rightarrow B + h\nu$$

processes make a significant contribution, the actinometer approach can be invalid. In addition, spatial and time resolved emission data (*118*) indicate that the electron distribution across the plasma is not uniform. As a result, normalized emission data depend on the region of the plasma observed. Therefore, a great deal of care must be maintained to ensure the accuracy of the data and its interpretation.

Of course, similar methods can be applied to monitoring emission from etch products excited in the discharge; however, if the etch process utilizes a stable molecular etchant or reactive species generated upstream from the wafer other methods must be employed. Recently, another study (*119*) has demonstrated an optical emission method for product monitoring which is not only applicable to those cases, but also is valuable in eliminating the effects of discharge variables in normal plasma reactors. The apparatus (Figure 26) employs a second discharge to excite emission from products downstream from the etch reaction zone. Although other downstream monitors are available (*120*), this method differs in that only a small sampling of the reactor effluent is admitted into the monitoring discharge. Therefore, provided that appropriate pressure control is maintained at the sampling site, product emission is totally independent of all process discharge variables and the parameters for the excitation discharge can be tailored to optimize sensitivity. In addition, by adding a known flow of pro-

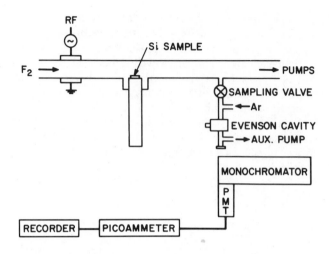

Figure 26. Experimental configuration for monitoring optical emission from etch products using an auxilliary discharge located downstream from the etch reaction zone. (Reproduced with permission from Ref. 119.)

duct to the effluent before the sampling point, absolute etch rates can be determined. Etch rate determinations on the order of 300Å/min per cm^2 of film area were demonstrated which could easily be reduced to a few Å/min with minor improvements. The primary disadvantage of downstream monitoring is that products with low volatility may condense on the walls thus creating errors and a severe background problem.

Another type of optical emission technique for product monitoring involves chemiluminescent reactions which are frequently obscured by electron-excited emission in most reactors. Although such reactions involving a direct etch product are rare, one very important one has been observed. This case offers sufficient advantages in end point detection that process reactors in which the active etchant is formed upstream from the wafer might be employed in critical situations where isotropic etching is tolerable. During the etching of Si with atomic fluorine (*43a,121*) generated upstream from the reaction zone, an intense visible emission (350-800nm) was observed above the wafer due to the chemiluminescent reaction sequence (Equations 18, 19) which produced electronically excited SiF$_3$. At high pressures (1-10 Torr) the intensity increased and was localized near the surface of the wafer. As a result, not only could the end point at an SiO$_2$/Si interface (Figure 27) be detected with high precision, but also the

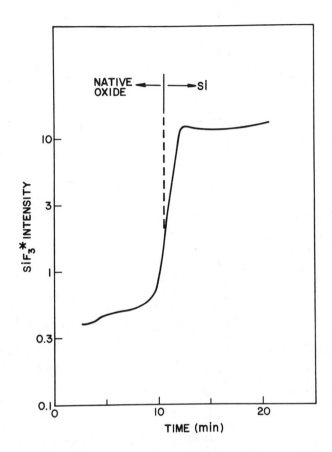

Figure 27. End point detection of a native oxide/silicon interface using chemiluminescence from SiF₃ formed by the reaction of the etch product SiF₂ with F and F₂. Etching was performed with F atoms downstream from at F₂ discharge.*

etching and end point of individual features on a wafer could be monitored visually. Such a monitoring technique in a production version of this reactor would permit the operator to inspect the progress of etching visually, make compensations for any loading effects, and ensure minimal overetching for any region on the surface of the wafer.

Optical emission methods have been used to obtain a great deal of information concerning the elementary processes in plasmas — how short-lived species are formed, how they react with substrates, and which discharge parameters are important in governing both. It would be impossible to enumerate all of the studies in the space allotted. Most of the citations given for Section 5.4 contain applications of optical emission to a variety of gases and films, and a more extensive bibliography can be obtained from a recent review (103). It is hoped that the limited discussion here has indicated some of the major concepts in using optical emission to unravel the miriad of elementary reactions that go on in a plasma reactor.

The use of optical emission for end point detection does not require the extensive understanding indicated above. All that is necessary is that a particular spectral feature change intensity at the interface between two films. Of course, to prevent or control overetching the change in intensity must be quite distinct. Both etchant and product emission have been used for end point detection during semiconductor and metal etching and plasma stripping of photoresists. Generally, product monitoring is superior since etchant monitoring requires a pronounced loading effect (38) for sensitive end point detection. Since the reactive species is typically present in a loaded reactor in large quantities, end points generally correspond to a small change in intensity on a large background. Product emission, on the other hand, can be used in an unloaded reactor, but problems may arise in a heavily loaded reactor. Since the etch rate increases as the loading effect diminishes, under some circumstances the net flux of product may not change significantly when the desired end point is reached. Thus, the best approach is to employ both product, and etchant, monitoring simultaneously. For VLSI applications where etch depths become comparable to feature separation, both methods have limited value in end point detection. Here, the total area (side wall + bottom) of material being etched can remain nearly constant even after the bottom has been reached and only undercutting is occurring. In such cases, optical reflection or emission monitoring of an etch product from an underlying film is obviously superior.

In practice, optical emission has proven to be a powerful end point detector. Although the apparatuses depicted in Figures 21 and 26 have indicated that monochromators are necessary, end point detection interference filters that select a predetermined wavelength and photodiode detectors can be used very effectively. Table V (109) summarizes some of the species and wavelengths that are useful for end point detection of several film materials.

5.6.c.2. Mass Spectrometry. Like optical emission spectroscopy, mass spectrometry offers the ability to fingerprint and identify individual species in a plasma discharge or products in the effluent from a plasma reactor. Its most common application is the latter, and a diagram for effluent monitoring by

Table V: Species and Emission Wavelengths
Useful for End Point Detection (*109*)

Film To Etch	Species Monitored	Wavelength (nm)	Reference
Resist	CO^*	297.7, 483.5, 519.5	*96a*
	OH^*	308.9	*96a*
	CO^*/OH^*	283	*96a*
	H_*	656.3	*96a,122*
Si, Poly-Si	F^*	704	*113,122,123*
	SiF^*	777	*124*
SiN, Si_3N_4	F^*	704	*123*
	CN^*	387	*123*
	N^*	674	*124*
Aluminum	$AlCl^*$	261.4	*125*
	Al^*	396	*109*
GaAs	As^*	278	*68*
	Ga^*	287.4	*68*

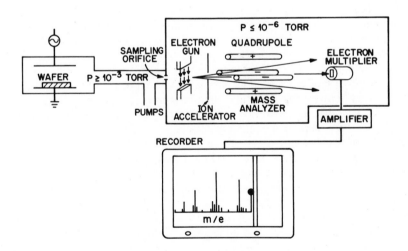

Figure 28.   Diagram for mass spectrometric monitoring of effluents from a plasma etching reactor.

mass spectrometry is shown in Figure 28. A small orifice is used to sample the process effluent, and electron impact is employed to form ions in much the same way as in the plasma reactor. The ions are accelerated into an electrostatic (shown here) or magnetic mass analyzer which separates the ions according to their mass-to-charge ratio (m/e) and focuses them onto an ion detector. Since the ions produced are generally of unit charge, a plot of ion intensity *vs.* m/e corresponds to a map of the masses of ions produced by the electron impact process. Pressures in the mass spectrometer are kept below 10^{-6} Torr to prevent ion-ion and ion-molecule reaction which would complicate mass analysis.

As noted earlier, electron impact reactions in the plasma produce a variety of species owing to processes such as

$$
\begin{aligned}
AB + e &\rightarrow AB^+ + 2e \\
&\rightarrow A^+ + B^- + e \\
&\rightarrow A^+ + B + 2e
\end{aligned}
\tag{30}
$$

Similar phenomena occur in the ion source of the mass spectrometer as well, and more often than not, considerable fragmentation occurs, leading to a rich spectrum of masses for each component of the sampled gas. For example, the mass spectrum of CF_3CF_2Cl (Figure 29) is rich in ions produced by the loss of F and Cl, along with fragments formed by the scission of the C-C bond. The fragmentation pattern provides an identifiable fingerprint of the molecule, even though as in this case, the parent peak (m/e=154) may be absent. The observed fragmentation pattern depends strongly on electron energies used in the ionization process. For optimum ionization efficiency, 50-100eV electrons, which produce high sensitivity and increased fragmentation, are employed. In such cases the parent ion is rarely observed, and in a complex mixture of gases such as plasma reactor effluent, the mass spectrum can be very complicated and difficult to interpret. To minimize fragmentation and simplify interpretation, electron energies can be reduced to 10-20eV; however, there is an inherent loss in sensitivity which may prohibit detection of important etch products. Therefore, the application of mass spectrometry to understanding plasma processes requires the experimenter to operate in both energy regimes. In this way, appropriate ions can be identified and spectrometer parameters optimized for detecting a desired species.

By identifying reactive species (i.e., F, CF_2, O, etc.) in the plasma using emission spectroscopy and by identifying the stable species (i.e., C_xF_y, COF_x, Si_xF_y) in the effluent with mass spectrometry, a great deal of insight into plasma reactions can be gained. For example, the development of the saturate-unsaturate theory (Section 5.4.b.) to elucidate the mechanism of

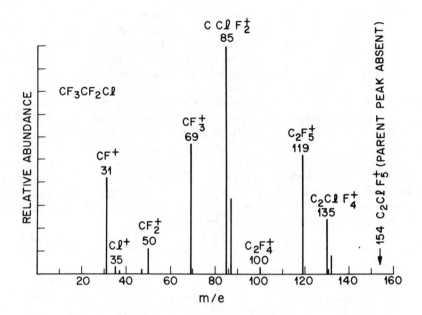

Figure 29. Mass spectrum of monochloropentafluoroethane (CF₃CF₂Cl).

radical production and assess the role of added oxidants in halocarbon plasmas relied heavily on such an approach. By comparing mass spectra of CF_3Cl with and without plasma excitation to those obtained in the presence of varying amounts of O_2, it was shown that the presence of the oligimers $(CF_2)_nClF$ in the mass spectrum of the effluent of CF_3Cl discharges correlated with the low F-atom content of the plasma (41). The addition of oxygen produced a marked decrease in the oligimer products and a concomitant increase in F atoms. As a result, plasma conditions conducive to etching or film deposition could be understood from a fundamental point of view.

Although effluent monitoring is most often used in practice, mass spectrometry can also be used to extract ions directly from the plasma (126). In this case the sampling orifice is in one of the plasma electrodes, and the ion accelerator extracts ions without using the electron-impact ionizer. In this way it is possible to identify ions present in the plasma that may not be observable by optical emission, and to gain insight into the energetics of the plasma and the role of ion-molecule reactions. Figure 30 shows the mass spectrum of positive ions extracted from a CF_4/O_2 discharge with a silicon target extraction electrode (*127*). Unfortunately, this study could not unambiguously assess the role of ion chemistry in the formation of the ions extracted. The dominant path appears to be electron-impact dissociation of CF_4 followed by radical-radical reactions leading to neutrals of the ions detected (produced by electron impact in the plasma before extraction).

Mass spectrometry is also extremely useful as a process monitor. Less sophisticated residual gas analyzers (RGA) operating on the principles of mass spectrometry are available for these purposes and for end point detection. For the etching of Si (*128-130*), poly-Si (*130*), silicon nitride (*130*), and SiO_2 (*129*), SiF_3^+ (m/e=85) has been shown to be effective for end-point detection. In addition, N^+ (m/e=14) is useful for nitride (*129,130*) in leak tight systems, while O^+ (m/e=16), CO^+ (m/e=44) and Si^+ (m/e=29) are useful for oxide (*131*). Because of the general nature of mass spectrometry as a diagnostic tool, it should be applicable to etching studies of metals and other semiconductor materials.

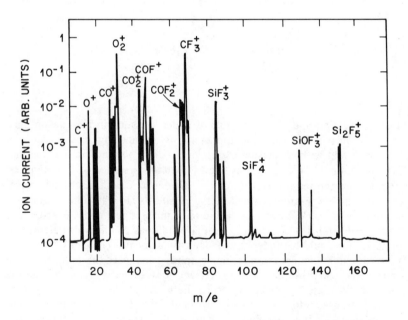

Figure 30. Mass analysis of ions extracted from a CF_4/O_2 discharge using a silicon target extraction electrode. (Reproduced with permission from Ref. 127.)

5.6.d. Other Diagnostic Techniques.   In the preceeding section a summary of the principal plasma diagnostics currently used was presented. Here we briefly summarize a few other techniques which have been either applied to gain a more fundamental understanding of plasma etch systems or to offer a novel approach to provide additional insight. Unfortunately, details cannot be given due to space limitations; however, it is hoped that the information and the literature citations given are sufficient to encourage the reader to explore these areas in greater depth.

5.6.d.1. Langmuir Probes.   A variety of probe techniques are available for plasma diagnostics (*132-134*). Langmuir (or electric) probes use one or more conducting wires (probes) inserted within a plasma to measure the electron and ion current drawn by the probe as the voltage applied to the probe is varied. Such information allows the determination of floating potential, plasma potential, electron density, electron temperature, and even electron energy distribution. Unfortunately, extraction of precise values or expressions for the latter three parameters can be complicated because of interpretational difficulties with the current-voltage data. In addition, because reactive gases are used for plasma etching, the probe material may be attacked or films may be deposited onto the probe, thereby introducing errors in current measurement.

A typical current-voltage characteristic of a probe is shown in Figure 31. Three basic segments can be identified in the I-V trace: (a) a region of electron (or negative particle) current where the probe is positive; (b) a region of positive ion current where the probe is negative; and (c) a transition region in which the current passes through zero (floating potential) and increases rapidly with increasing positive potentials. From the variation of current with voltage in the transition region, the electron temperature may be determined. Also, the plasma potential occurs at the "knee" in the probe characteristic as the electron current begins to saturate.

5.6.d.2. Surface Probes.   Since the process of plasma etching involves the surface of a film, any probe capable of monitoring surface species would be most valuable in understanding the types of reactions that ultimately lead to the gasification of a film. Two that have been applied to plasma etching are X-ray photoelectron spectroscopy (ESCA or XPS) and Auger spectroscopy. In photoelectron spectroscopy (*135*) an X-ray (ESCA or XPS) or vacuum ultraviolet beam (PES) is directed at the film surface. These radiations liberate electrons that are energy analyzed to display characteristic spectra of kinetic energies. The difference between the incident photon energy and the detected electron energy gives the binding energy of the ejected electron which is characteristic of the elemental and molecular structure of the surface. Thus, by determining the carbon, oxygen, fluorine content of a silicon surface (*136*) it is possible to identify adsorbed and bound surface species and gain insight into the mechanism of the etch process. Auger spectroscopy (*137*) is similar to ESCA and PES in that surface analysis is based on energy analysis of electrons ejected; however, in Auger spectroscopy the excitation source is an electron beam. Although similar information (*136*) is obtained by both methods, Auger spectroscopy is capable of analyzing smaller areas of the surface. Unfortunately, neither method is applicable to analyses during etching and care must be exercised to avoid contamination when transferring samples to the analysis chamber. Generally, only strongly chemisorbed species are detectable since the energy of the incident beam (particularly in Auger studies) is sufficient to desorb weakly bound reactants and products.

5.6.d.3. Gas Phase Probes. Two novel techniques that should prove to be very valuable diagnostics in the arsenal of the plasma scientist are electron paramagnetic resonance (EPR) spectroscopy and laser-induced fluorescence (LIF) spectroscopy (*138*). Both offer the advantage of being able to sensitively detect transient gaseous species such as radicals, atoms and ions in a nonintrusive way. In EPR spectroscopy a magnetic field is used to remove the electron spin and/or orbital angular momenta degeneracies of species

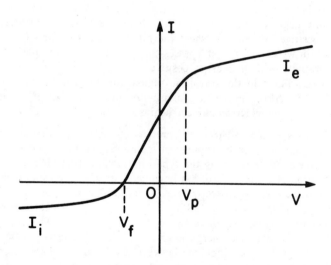

Figure 31. Current-voltage characteristics of an electric probe. The ion and electron current are I_i and I_e, and the floating and plasma potentials are V_f and V_p, respectively.

(radicals, atoms) which exhibit a magnetic moment. The applied field is varied until the separations of the magnetically tunable energy levels of the radical species are in resonance with microwave electromagnetic radiation. When this condition is satisfied, the absorption of microwave energy is detected. Although it must be used on the effluent of most reactors, it is possible to detect in a quantitative fashion the number density of important ground state species such as F, O, Cl, and Br. Recalling that optical emission detects only *excited* states (radiative lifetimes $\approx 10^{-9}$ sec) of species, such an ability offers considerable advantage since ground state species are the most active in the etch process. EPR has been applied to monitoring O and F atoms in photoresist stripping (95) and should find more extensive application to plasma processing in general.

LIF should become a widely used tool as new laser techniques are developed. A truly nonintrusive technique, LIF provides the ability to excite ground state species and monitor emission arising from the excitation. Again one has a ground state probe; however, there is an even greater advantage over optical emission excited by the plasma discharge. Since one has control over the excitation wavelength as well as the emission wavelength monitored, very high selectivity and greater simplicity in the spectra results. Preliminary results on the application of LIF to Cl_2 (72), CCl_4 (139), and fluorocarbon (140) plasmas have indicated that gated detection provides excellent discrimination against the glow discharge background with extremely high sensitivity. In addition, the ability to excite fluorescence from atomic species via two-photon excitation has been demonstrated (141) and promises the ability to monitor the concentration of ground state atomic species without the necessity of vacuum ultraviolet sources and optics. Coupled with the high spatial and temporal resolution of the technique, LIF should provide new information to increase our understanding of plasma etching.

5.7. Other Dry Etch Techniques

Dry etching techniques permit etch processes to be carried out in various modes. These can be described as purely chemical, purely physical, and a mixture of chemical and physical. With plasma etching and RIE, we have concentrated on chemical and ion assisted processes. In this section, etching methods that depend either solely or primarily on physical processes (momentum transfer) will be discussed briefly.

5.7.a. Ion Milling. Conceptually, the simplest momentum transfer or sputtering (142) process is carried out by immersing the material to be etched in a glow discharge, and imposing either externally or via electrode area ratios, a negative bias on the material. In this technique, a chemically inert gas such as argon is generally used. Positive argon ions are generated in the glow discharge and are accelerated through the plasma sheath into the negatively charged substrate. When they strike the substrate, they transfer their momentum to the material. If the ion energy is sufficient, substrate atoms, molecules, and ions are ejected. In this way, etching is achieved.

In ion milling systems (*143-145*) a confined plasma is used to generate ions. A set of grids used for confinement is biased so that an ion beam can be extracted from the source. This beam is then directed to the substrate surface, where sputter etching or ion milling takes place.

A number of advantages exist for ion milling compared to plasma etching or RIE. Because of the collimated beam of ions, essentially vertical profiles are possible. Also, profile tapering can be achieved by tilting the substrate relative to the ion beam. In addition, ion milling is performed at pressures at least 100 times lower than those used in plasma etching or RIE. Therefore, redeposition of sputtered material is reduced.

Unfortunately, since ion milling is a purely physical process, selectivity is generally poor. Indeed, selectivity in such systems depends almost exclusively on differences in sputter yield between materials. Finally, since the etch products are not volatile, redeposition and trenching can be serious limitations (*146*).

5.7.b. Reactive Ion Beam Etching. In order to enhance etch selectivity and to increase etch rates, reactive ion beam etching (RIBE) has been developed (*147-149*). This technique is carried out in an manner analogous to ion beam milling, except that a confined plasma is established using a chemically reactive gas. Thus, a chemical component is imparted to a primarily physical process. In principle, since the substrates are not immersed in a plasma, parameter control is greatly simplified. However, the ultimate selectivities achievable are not yet clear. Further, neutral species may play a role in RIBE (*148*). If this is correct, parameter control may have limitations silimar to those encountered in plasma or RIE techniques.

5.8. Safety Considerations

As in any process that uses chemicals and electronic and mechanical equipment, a concerted safety effort is required in plasma etching. Proper shielding of reactors and power supplies to minimize operator exposure to rf radiation is imperative. An exposure level below 1 mW/cm^2 has been suggested as a safe operating point (*150*).

The handling of most of the chlorinated and brominated vapors used for plasma etching requires particular caution. For instance, carbon tetrachloride is toxic and a suspected carcinogen, while chlorine is toxic and highly corrosive. Often unappreciated, however, is the fact that products of plasma reactions may be more toxic and corrosive than the reactants. For example, a CCl_4 plasma generates C_2Cl_6 (a carcinogen), Cl_2, and various unidentified chlorinated organic compounds (*151*). If water vapor and oxygen are present in the plasma atmosphere, phosgene ($COCl_2$) and HCl are formed (*151*). Polymer deposits that form during etching often trap chlorocarbon vapors which can be released or can hydrolyze when the chamber is exposed to the atmosphere. Also, etch products such as $AlCl_3$ are strong Lewis acids and hydrolyze readily to form HCl.

Seemingly innocuous gases such as CF_4 can form COF_2 and HF if oxygen and water vapor are present in the plasma atmosphere. Carbonyl difluoride (COF_2) can subsequently hydrolyze to form HF when exposed to atmospheric conditions. Therefore, extensive system purging should be carried out so that etch gases or their products are removed prior to opening the chamber.

Care must also be exercised when cleaning reactor parts and especially when changing pump oil. The vapors formed or passed through the plasma reactor are dissolved in most oils. Thus, inhalation of vapors and contact of oil with the skin must be avoided during oil changes and pump maintenance.

Fluorine- and particularly chlorine-based fragments generated in the glow discharge can remove a hydrogen atom from hydrocarbon-based oils, thus initiating polymerization (*151*). Similar reactions occur with Lewis acids such as $AlCl_3$ and BCl_3. Oils composed of perfluoropolyethers appear to be resistant to many of the reactions undergone by silicone and hydrocarbon-based oils, at least at temperatures below $100°C$ (*152*).

Finally, oxidation of boron-, silicon-, and aluminum-containing materials results in oxide particles. Along with polymers in the oil, these particles can plug lubrication ducts, and cause pump failure (*153*). An oil filter is often used to remove such particles, thereby extending pump life and increasing the time

Acknowledgements. The authors would like to thank D. A. Danner and C. C. Tang for critically reading the manuscript. Also, D. W. H. acknowledges support of his plasma etching studies by the National Science Foundation under Grant ECS-8021508.

Literature Cited

1. Kern, W.; Deckert, C. A. *in* "Thin Film Processes"; Vossen, J. L.; Kern, W.; Eds.; Academic: New York, 1978, p. 401.
2. Deckert, C. A.; Peters, D. A., *Kodak Microelectronics Seminar Proceedings*, October, 1977, p. 13.
3. "The Application of Plasmas to Chemical Processing", Baddour, R. F.; Timmins, R. S., Eds.; MIT Press, Cambridge, MA. 1967.
4. "Chemical Reactions in Electrical Discharges," Blaustein, B. C., Ed.; American Chemical Society, Advances in Chemistry Series No. 80, 1969.

5. "Techniques and Applications of Plasma Chemistry", Hollahan, J. R.; Bell, A. T., Eds., Wiley: New York, 1974.

6. "Reactions Under Plasma Conditions", Venugopalan, M., Ed., Wiley Interscience, 1971.

7. Irving, S. M., *Kodak Photoresist Seminar Proceedings,*" Vol. II, May 20-21, 1968, p. 26.

8. Irving, S. M.; Lemons, K. E.,; Bobos, G. E., U.S. Pat. No. 3615956, October 26, 1971.

9. Bell, A. T., in Ref. 5, p. 1.

10. Kay, E.; Coburn, J.; Dilks, A. in "Plasma Chemistry III," Vol. 94 of "Topics in Current Chemistry", Springer-Verlag, 1980, p. 1.

11. Melliar-Smith, C. M.; Mogab, C. J. *in* "Thin Film Processes"; Vossen, J. L.; Kern, W., Academic Press: New York, 1978, p. 497.

12. Vossen, J. L., *J. Electrochem. Soc.,* 1979, *126*, 319.

13. Chapman, B. N., "Glow Discharge Processes"; Wiley: New York, 1980.

14. Butler, H. S.; Kino, G. S., *Phys. Fluids,* 1963, *6*, 1346.

15. Anderson, G. S.; Mayer, W. N.; Wehner, G. K., *J. Appl. Phys,* 1966, *37*, 574.

16. Ephrath, L. M.; DiMaria, D. J., *Solid State Technol.,* 1981, *24*(4), 182.

17. Koenig, H. R.; Maissel, L. I., *IBM J. Res. Dev.,* 1970, *14*, 276.

18. Coburn, J. W. and Kay, E., *J. Appl. Phys.,* 1972, *43*, 4965.

19. Bruce, R. H., *J. Appl. Phys.,* 1981, *52*, 7064.

20. Winters, H. F. *in* "Plasma Chemistry III" Vol. 94 of "Topics in Current Chemistry", Springer-Verlag, 1980, p. 69.

21. Bell, A. T., *Solid State Technol.,* 1978, *21*(4), 89.

22. McDaniel, E. W., "Collision Phenomena in Ionized Gases"; Wiley: New York, 1964.

23. Massey, H. S. W.; Burhop, E. H. S.; Gilbody, H. B., "Electronic and Ionic Impact Phenomena"; Oxford Press: New York, 1971.

24. Rutscher, A., "Progress in Electron Kinetics of Low Pressure Discharges and Related Phenomena", presented at the 13th International Congress on Phenomena in Ionized Gases, East Berlin, GDR, September, 1977.

25. McDaniel, E. W.; Cermak, V.; Dalgarno, A.; Ferguson, E. E.; Friedman, L., "Ion-Molecule Reactions": Wiley: New York, 1970.

26. Kondratiev, V. N. "Chemical Kinetics of Gas Reactions", Addison Wesley: Reading, Massachusetts, 1964.

27. Carter, G.; Colligan, J. S., "Ion Bombardment of Solids", Elsevier: New York, 1969.

28. Gerlach-Meyer, U.; Coburn, J. W.; Kay, E., *Surface Sci.,* 1981, *103*, 177.

29. Donnelly, V. M.; Flamm, D. L., *Solid State Technol.*, 1981, *24*(4), 161.
30. Coburn, J. W. *in Proceedings of the Tutorial Symposium on Semiconductor Technology*, Doane, D. A.; Fraser, D. B.; Hess, D. W., Eds., The Electrochemical Society, Inc., 1982.
31. Coburn, J. W.; Winters, H. F., *J. Appl. Phys.*, 1979, *50*, 3189.
32. *Ref.* 10, p. 6.
33. Bondur, J. A., *J. Vac. Sci. Technol.*, 1978, *13*, 1023.
34. Lehmann, H. W.; Widmer, R., *J. Vac. Sci. Technol.*, 1978, *15*, 319.
35. Horiike, Y.; Shibagaki, M. *in* "Semiconductor Silicon 1977"; Ruff, H. R.; Sirtl, E., Eds., The Electrochemical Society, 1977, p. 1071.
36. Flamm, D. L.; Donnelly, V. M.; Mucha, J. A., *J. Appl. Phys.*, 1981, *52*, 3633.
37. Flamm, D. L.; Mogab, C. J.; Sklaver, E. R., *J. Appl. Phys.*, 1979, *50*, 6211.
38. Mogab, C. J., *J. Electrochem. Soc.*, 1977, *124*, 1262.
39. Flamm, D. L.; Wang, D. N. K.; Maydan, D., *J. Electrochem. Society,* 1982, *129*, 2755.
40. Coburn, J. W.; Kay, E., *IBM J. Res. Develop.,* 1979, *23*, 33.
41. Flamm, D. L., *Plasma Chem. Plasma Processing,* 1981, *1*, 37.
42. Mogab, C. J.; Adams, A. C.; Flamm, D. L., *J. Appl. Phys.*, 1979, *49*, 3796.
43. (a) Donnelly, V. M.; Flamm, D. L., *J. Appl. Phys.*, 1980, *51*, 5273. (b) Mucha, J. A.; Donnelly, V. M.; Flamm, D. L.; Webb, L. M., *J. Phys. Chem.*, 1981, *85*, 3529. (c) Mucha, J. A.; Flamm, D. L.; Donnelly, V. M., *J. Appl. Phys.*, 1982, *53*, 4553.
44. (a) Chuang, T. J., *J. Appl. Phys.*, 1980, *51*, 2614. (b) Chuang, T. J., *Phys. Rev. Lett.*, 1979, *42*, 815.
45. Vasile, M. J.; Stevie, F. A., *J. Appl. Phys.*, 1982, *53*, 3799.
46. van de Ven, E. P. G. T.; Zijlstra, P. A., *Electrochem. Soc. Extended Abstracts*, May 1980, *80-1*, 253.
47. Horwath, R.; Zarowin, C. B.; Rosenberg, R., *Electrochem Soc. Extended Abstracts*, May 1980, *80-1*, 294.
48. Boyd, H.; Tang, M. S., *Solid State Technol.*, 1979, *25*(4), 133.
49. Eisele, K. M., *J. Electrochem. Soc.*, 1981, *128*, 123.
50. d'Agostino, R.; Flamm, D. L., *J. Appl. Phys.*, 1981, *52*, 162.
51. Flamm, D. L.; Cowen, P. L.; Golovchenko, J. A., *J. Vac. Sci. Tech*, 1980, *17*, 1341.
52. Bruce, R. H., *Solid State Technol.*, 1981, *24*(10), 64; *J. Appl. Phys.,* 1981, *52*, 7064.
53. Mogab, C. J.; Levenstein, H. J., *J. Vac. Sci. Technol.*, 1980, *17*, 721.
54. Adams, A. C.; Capio, C. D., *J. Electrochem Soc.*, 1981, *128*, 366.
55. Heinecke, R. A. H., *Solid State Electron.*, 1975, *18*, 1146; 1976, **19**, 1039.

56. Kirmse, W. "Carbene Chemistry", Academic: New York, 1971.
57. Gray, P.; Herod, A. A.; Jones, A., *Chem. Revs.*, 1971, *71*, 247.
58. Coburn, J. W.; Winters, H. F., *J. Vac. Sci. Technol.*, 1979, *16*, 391.
59. Coburn, J. W.; Winters, H. F., *Solid State Technol.*, 1979, *22*(4), 117.
60. Matsuo, S., *J. Vac. Sci. Technol.*, 1980, *17*, 587.
61. Mayer, T. M., *J. Electronic Materials*, 1980, *9*, 513.
62. Toyoda, H.; Komiya, H.; Itakura, H., *J. Electronic Materials*, 1980, *9*, 569.
63. Matsuo, S., *Jap. J. Appl. Phys.*, 1978, *17*, 235.
64. Reinberg, A. R.; Dalle Ave, J.; Steinberg, G.; and Bruce, R., *Electrochem. Soc. Extended Abstracts*, October 1981, *81-2*, 669.
65. Jacob, A., *Solid State Technol.*, 1978, *21*(4), 95.
66. (a) Donnelly, V. M.; Flamm, D. L.; Tu, C. W.; Ibbotson, D. E., *J. Electrochem. Soc.*, 1982, *129*, 2533. (b) Ibbotson, D. E.; Flamm, D. L.; Donnelly, V. M.; Duncan, B. S., *J. Vac. Sci. Technol.*, 1982, *20*, 489. (c) Ibbotson, D. E.; Donnelly, V. M.; Flamm, D. L.; *Extended Abstracts Electrochem. Soc.*, October 1981, *81-2*, 650.
67. Smolinsky, G.; Chang, R. P.; Mayer, T. M., *J. Vac. Sci. Technol.*, 1981, *18*, 12.
68. Klinger, R. E.; Greene, J. E., *Appl. Phys. Lett.*, 1981, *38*, 620.
69. Hu, E. L.; Howard, R. E., *Appl. Phys. Lett.*, 1980, *37*, 1022.
70. Coldren, L. A.; Rentschler, J. A., *J. Vac. Sci. Technol.*, 1981, *19*, 225.
71. Bösch, M. A.; Coldren, L. A.; Good, E., *Appl. Phys. Lett.*, 1981, *38*, 264.
72. Donnelly, V. M.; Flamm, D. L.; Collins, G. J.; (a) *Extended Abstracts Electrochemical Society*, October 1982, *81-2*, 621, (b) *J. Vac. Sci. Technol.*, 1982, *21*, 817.
73. Poulsen, R. G., *J. Vac. Sci. Technol.*, 1977, *14*, 266.
74. Hess, D. W., *Plasma Chem. Plasma Processing*, 1982, *2*, 141.
75. Herb, G. K.; Porter, R. A.; Cruzan, P. D.; Agraz-Guerena, J.; Soller, B. R., *Electrochem. Soc. Extended Abstracts,* October 1981, *81-2*, 710.
76. Legat, W. H.; Schilling, H., *Electrochem. Soc. Extended Abstracts*, October 1975, *75-2*, 336.
77. Mogab, C. J.; Shankoff, T. A., *J. Electrochem. Soc.*, 1977, *124*, 1766.
78. Nakata, H.; Nishioka, K.; Abe, H., *J. Vac. Sci. Technol.*, 1980, *17*, 1351.
79. Harada, T.; Gamo, K.; Namba, S., *Jap. J. Appl. Phys.*, 1981, *20*, 259.
80. Randall, J. N.; Wolfe, J. C., *Appl. Phys. Lett.,* 1981, *39*, 742.
81. Tokunaga, K.; Redeker, F. C.; Danner, D. A.; Hess, D. W., *J. Electrochem. Soc.*, 1981, *128*, 851.

82. Winkler, U.; Schmidt, F.; Hoffman, N., Recent News Paper No. 605, presented at the St. Louis Meeting of the Electrochemical Society, May 11-16, 1980.

83. Schaible, P. M.; Metzger, W. C.; Anderson, J. P., *J. Vac. Sci. Technol.*, 1978, *15*, 334.

84. Poulsen, R. G.; Nentwich, H.; Ingrey, S., *Proc. of the International Electron Devices Meeting, Washington, D.C.*, 1976, p. 205.

85. Lee, W. Y.; Eldridge, J. M.; Schwartz, G. D., *J. Appl. Phys.*, 1981, *52*, 2994.

86. Fok, Y. T., *Electrochem. Soc. Extended Abstracts*, May 1980, *80-1*, 301.

87. Parry, P. D.; Rodde, A. F., *Solid State Technol.*, 1979, *22*(4), 125.

88. Lehmann, H. W.; Widmer, R., *Appl. Phys. Lett.*, 1978, *32*, 163.

89. Bondur, J. A., *J. Electrochem. Soc.*, 1979, *126*, 226.

90. Jinno, K., *Jap. J. Appl. Phys.*, 1978, *17*, 1283.

91. Matsuo, S.; Takehara, Y., *Jap. J. Appl. Phys.*, 1977, *16* 175.

92. Ephrath, L. M., *J. Electrochem Soc.*, 1979, *126*, 1419.

93. Toyoda, H.; Komiya, H.; Itakura, H., *J. Electronic Materials*, 1980, *9*, 569.

94. Taylor, G. N.; Wolf, T. M., *Polym. Eng. and Sci.*, 1980, *20*, 1086.

95. Benson, B. W.; Cook, J. M., unpublished results (1980).

96. *See*, for example, (a) Degenkolb, E. O.; Mogab, C. J.; Goldrick, M. R.; Griffiths, J. E., *Appl. Spectrosc.*, 1976, *30*, 520; (b) Battey, J. F., *IEEE Trans. Electron. Dev.*, 1977, *ED-24*, 140; (c) Stafford, B. B.; Gorin, G. J., *Solid State Technol.*, 1977, *20*(9), 51; and (d) Szekeres, A.; Kirov, K.; Alexandrova, S., *Phys. Stat. Sol. A*, 1981, *63*, 371.

97. Niebauer, D. A., *Electron. Packaging Prod.*, p. 153 (1981); Kegel, B., *Circuits Manuf.*, 1981, *21*, 27.

98. (a) Taylor, G. N.; Wolf, T. M., *J. Electrochem. Soc.*, 1980, *127*, 2665; (b) Taylor, G. N.; Wolf, T. M.; Moran, J. M., *J. Vac. Sci. Technol.*, 1981, *19*, 872.

99. Calvert, J. G.; Pitts, J. N., Jr., "Photochemistry", Wiley: New York, 1967.

100. Howard, R. E.; Prober, D. E., "Nanometer-Scale Fabrication Techniques" *in* "VLSI Electronics-Microstructure Science", Vol. 5, Einspruch, N. G., ed., 1982.

101. Brandes, R. G.; Dudley, R. H., *J. Electrochem. Soc.* 1973, *120*, 140.

102. Robb, F., *Semiconductor International*, 1979, *2*(11), pp. 60-65.

103. Flamm, D. L.; Donnelly, V. M., *Plasma Chem. and Plasma Proc.* 1981, *1*, 317.

104. (a) Tu, Y. Y.; Chuang, T. J.; Winters, H. F., *Phys. Rev. B.*, 1981, *23*, 823; (b) Gerlach-Meyer, U.; Coburn, J. W.; Kay, E., *Surf. Sci.*, 1981, *103*, 177.

105. Adams, A. C.; Capio, C. D., *J. Electrochem. Soc.*, 1981, *128*, 366.
106. Bruce, R.; Flamm, D. L.; Donnelly, V. M.; Duncan, B. S., *J. Appl. Phys.*, in press (1983).
107. Wang, D. N. K.; Maydan, D., unpublished results (1979).
108. Coburn, J. W., *Plasma Chem. and Plasma Proc.*, 1982, *2*, 1.
109. Marcoux, P. J.; Foo, P. W., *Solid State Technol.*, 1981, *24*(4), 115.
110. Busta, H. H.; Lajos, R. E.; Kiewit, D. A., *Solid State Technol.*, 1979, *22*(2), 61.
111. Kleinknecht, H. P.; Meier, H., *J. Electrochem. Soc.*, 1978, *125*, 798.
112. Berning, P. H. *in* "Physics of Thin Films; Advances in Research and Development"; Hass, G., Ed.; Academic: New York, 1963; Vol. 1 pp. 69-120.
113. Harshbarger, W. R.; Porter, R. A.; Miller, T. A.; Norton, P., *J. Appl. Spectrosc.*, 1977, *31*, 201.
114. *See*, for example, Pearse, R. W. B.; Gaydon, A. G. "The Identification of Molecular Spectra" Chapman & Hall, London, 1965; 3rd Ed. for molecular spectra and Zaidel, A. N.; Prokof'ev, V. K.; Raiskii, S. M.;, Slavnyi, V. A.; Shreider, E. Ya., "Tables of Spectral Lines"; Plenum: New York, 1970; for atomic spectra.
115. Flamm, D. L.; Donnelly, V. M.; Ibbottson, D. E., *J. Vac. Sci., Technol.*, 1983, in press.
116. Coburn, J. W.; Chen, M., *J. Appl. Phys.* 1980, *51*, 3134.
117. d'Agostino, R.; Cramarossa, F.; DeBenedictis, S.; Ferraro, G., *J. Appl. Phys.* 1981, *52*, 1259.
118. Donnelly, V. M., private communication (1982).
119. Danner, D. A.; Mucha, J. A.; Flamm, D. L., *J. Electrochem. Soc.*, 1983, in press.
120. Singleton, M. J., *Solid State Technol.*, 1981, *24*(4), 132.
121. (a) Beenakker, C. I. M.; van Dommelen, J. H. J.; Dieleman, J., *Extended Abstracts, 157th Meeting of the Electrochemical Society*, 1980, *80-1*, 330. (b) Zijlstra, P. A.; Beenakker, C. I. M., *Appl. Spectrosc.*, 1981, *35*, 413.
122. Poulsen, R. G.; Smith, G. M., *Semiconductor Silicon*, 1971, *77*, 1058.
123. (a) Harshbarger, W. R.; Porter, R. A., *Solid State Technol.*, 1978, *21*(4), 99. (b) Harshbarger, W. R.; Porter, R. A.; Norton, P., *J. Electronic Materials*, 1978, *7*, 429.
124. Hirobe, K.; Tsuchimoto, T., *J. Electrochem. Soc.*, 1980, *127*, 234.
125. (a) Curtis, B. J., *Solid State Technol.*, 1980, *23*(4), 129. (b) Curtis, B. J.; Brunner, H. J., *J. Electrochem. Soc.*, 1978, *125*, 829.
126. (a) Coburn, J. W., *Rev. Sci. Inst.*, 1970, *41*, 1219. (b) Kay, E.; Coburn, J. W.; Kruppa, G., *Vide*, 1976, *183*, 89.
127. Winters, H. F.; Coburn, J. W.; Kay, E., *J. Appl. Phys.*, 1977, *48*, 4973.

128. Bunyard, G. B.; Raby, B. A., *Solid State Technol.*, 1977, *20*(12), 53.
129. Brown, H. L.; Bunyard, G. B.; Lin, K. C., *Solid State Technol.*, 1978, *21*(7), 35.
130. Raby, B. A., *J. Vac. Sci. Technol.*, 1978, *15*, 205.
131. Oshima, M., *Jap. J. Appl. Phys.*, 1978, *17*, 579.
132. Chen, F. F. *in* "Plasma Diagnostic Techniques", Huddlestone, R. H.; Leonard, S. L., Eds., Academic: New York, 1965, Ch. 4.
133. Swift, J. D.; Schwar, M. J. R., "Electric Probes for Plasma Diagnostics", Elsevier, New York, 1969.
134. Clements, R. M., *J. Vac.Sci. Technol.*, 1978, *15*, 193.
135. (a) Siegbahn, K., et. al. "ESCA Applied to Free Molecules", North-Holland: Amsterdam, 1970. (b) Turner, D. W.; Baker, C.; Baker, A. D.; Brundle, C. R. "Molecular Photoelectron Spectroscopy", Wiley: New York, 1970.
136. (a) Coburn, J. W.; Winters, H. F.; Chuang, T. J., *J. Appl. Phys.*, 1977, *48*, 3532. (b) Chuang, T. J., *J. Appl. Phys.*, 1980, *51*, 2614.
137. Carlson, T. A., "Photoelectron and Auger Spectroscopy"; Plenum: New York, 1970.
138. *See*, for example, Miller, T. A., *Plasma Chem. and Plasma Processing*, 1981, *1*, 3, and references contained therein.
139. Gottscho, R. A.; Burton, R. H.; Davis, G. P., *J. Chem. Phys.*, 1982, *77*, 5298.
140. Hargis, P. J., Jr.; Kusher, M. J., *Appl. Phys. Lett.*, 1982, *40*, 779.
141. Heaven, M.; Miller, T. A.; Freeman, R.; White, J.; Bokor, J., *Chem. Phys. Lett.*, 1982, *86*, 458.
142. *See*, for example, Part II of "Thin Film Process"; Bossen, J. L.; Kern, W., Eds., Academic: New York, 1978, pp. 12-208.
143. Bollinger, D.; Fink, R., *Solid State Technol.*, 1980, *23*(11), 79.
144. Hawkins, D. T., *J. Vac. Sci. Technol.*, 1979, *16*, 1051.
145. Lee, R. E., *J. Vac. Sci. Technol.*, 1979, *16*, 164.
146. Mogab, C. J. *in* "Physics of Dielectric Solids, 1980"; Goodman, C. H. L., Ed., The Institute of Physics, 1980, p. 37.
147. Meusemann, B., *J. Vac. Sci. Technol.*, 1979, *16*, 1886.
148. Downey, D. F.; Bottoms, W. R.; Hanley, P. R., *Solid State Technol.*, 1981, *24*(2), 121.
149. Mayer, T. M.; Barker, R. A., *J. Electrochem. Soc.*, 1982, *129*, 585.
150. Chapman, B.; Nowak, M., *Semiconductor International*, 1980, *3*(11), 139.
151. Hess, D. W., *Solid State Technol.*, 1981, *24*(4), 189.
152. O'Hanlon, J. F., *Solid State Technol.*, 1981, *24*(10), 86.
153. Connock, P., ETE Advanced Plasma Technology Seminar, January 1982.

RECEIVED February 4, 1983

6

Multi-Layer Resist Systems

B. J. Lin

IBM T. J. Watson Research Center
Yorktown Heights, New York 10598

0097-6156/83/0219-0287$15.60/0
© 1983 American Chemical Society

6.1 Introduction

The requirements on resist systems for modern IC fabrication are becoming more and more difficult to meet as the minimum feature size is reduced to the micrometer and submicrometer regime. As more exotic exposure sources are used, the available exposure intensity becomes limited, and the sensitivity of resist systems has to be improved to maintain a reasonable throughput. Furthermore, for smaller features, the vertical dimensions do not shrink as rapidly as the lateral dimensions, therefore, the height-to-width aspect ratio of resist images has to be higher. For example, a resist thickness of 1.5 - 2.0 μm is required for 1 - μm features. It reduces to 1-1.5 μm only when the feature size is reduced to 0.5 μm. Despite the aspect ratio requirement, maintaining linewidth control over high steps and at different parts of the wafer topography also becomes more difficult as the feature size becomes smaller. Many other problems associated with the special characteristics of each lithographic imaging system also arise as the minimum feature size is reduced. For example, optical interference effects caused by reflection from the topographic features becomes a significant contribution to linewidth variation. Proximity effects caused by backscattered electrons make high-aspect-ratio e-beam imaging impossible. Most of these problems can be alleviated by using a multi-layer resist (MLR) system.

As shown in Figure 1, a thick resist layer is first applied over topography, to provide a planar surface upon which a thin imaging resist layer can coat uniformly. After the imaging layer is delineated, it is used as a blanket exposure mask or etch mask to delineate the planarizing layer. Because this mask always conforms to the planarizing layer and is portable with the wafer, it can be called a portable conformable mask (PCM). Many times, a third layer is used between the imaging and the planarizing layers as the PCM. This becomes a three-layer system in which the imaging and masking functions of the top layer are now separated. For special applications, more than three layers may be required. The thickness of the planarizing layer is usually between 1 and 4 μm, depending on the wafer topography and other considerations, as will be discussed at appropriate places in the rest of this chapter. The thickness of the imaging layer and the isolation layer is 0.3 - 0.5 μm and 0.05 - 0.2 μm respectively and will also be discussed later in greater detail.

MLR systems can be classified by the number of layers and the function of the PCM, e.g., two-layer RIE PCM. Some authors prefer to use "bi-level," "tri-level," "double-layer," "triple-layer," or any other permutation of these words to denote the number of layers. The word "layer" and simple numerical adjectives are used here, because "level" may cause confusion because of its association with masking levels and because simple numerical adjectives pronounce better when the number of layers exceeds three.

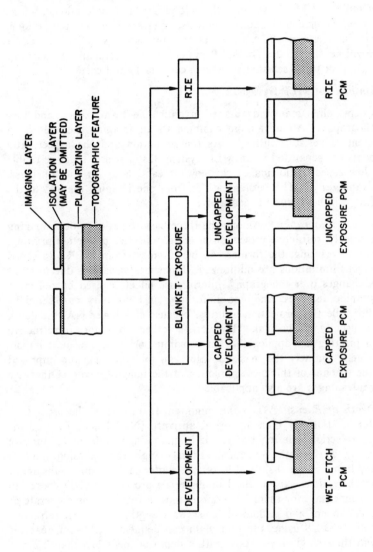

Figure 1.   Multi-layer resist systems.

In this chapter, the motivations to adopt MLR systems for optical e-beam, x-ray, and ion-beam lithographic systems will be given, followed by a survey of published MLR systems. Specific practical considerations such as planarization, pinhole and additive defects, interfacial layer, etch residue, film stress, interference effects, spectral transmission, inspection and resist stripping will be discussed. The MLR systems will be compared in terms of resolution, aspect ratio, sensitivity, process complexity and cost.

6.2 Motivations for MLR Systems

MLR systems offer many advantages in optical, e-beam, x-ray, and ion-beam lithography. An advantage common to all imaging methods is in enhancement of resist sensitivity. As the resolution and the aspect ratio requirements are separated in an MLR system, faster resists that are usable only for low aspect ratio images can now be candidates for the top layer. Other advantages of MLR systems differ from one imaging method to the other. They will be discussed separately.

6.2.a Optical Lithography. Among the three basic types of optical imaging schemes, namely, refractive projection printing, reflective projection printing, and proximity printing, the former can benefit most from MLR. In short, depth of focus limitations are minimized. Linewidth variations due to resist thickness changes over topography, interference effects caused by reflection off topographic features, and non-uniformity of reflectivity can be eliminated. For reflective projection printing, because of a lower NA, the depth of focus is usually less critical. However, the other advantages in refractive projection printing are applicable. For proximity printing, the most important parameter, namely the mask-to-wafer gap tolerance, can be improved only by the amount of thickness reduction of the imaging layer. Otherwise the other advantages are also applicable.

6.2.a.1 Depth of Focus. When the minimum feature is of the order of a micrometer, a relatively high numerical aperture (NA) has to be used for refractive projection printing because it is increasingly difficult to improve resolution by wavelength reduction. Even though the exposing light is bandpass filtered to a bandwidth of the order of 10 nm, substantial chromatic aberrations exist if the lens materials are not properly chosen to mutually compensate for them. Therefore, most refractive lenses operate at 405 and/or 436 nm and seldom at lower wavelength. A typical refractive lens with NA = 0.28 using 436 - nm light can delineate 1.25 - μm features in a 1 - μm thick single layer resist with a depth of focus less than 2 μm if satisfactory linewidth control and a realistic illumination uniformity are considered. When a 2 - μm resist thickness is required, its effective optical thickness is 1.25 μm after dividing by a typical resist refractive index of 1.6. There is only a 0.75 - μm margin of error for focus tolerance and wafer unevenness. On the other hand, a 0.5 - μm imaging layer in an MLR system leaves 1.69 μm of margin as shown in Figure 2.

6.2.a.2 Linewidth Variations due to Resist Thickness. As the imaging light propagates through the resist, attenuation and multiple reflections occur. Since energy in the light beam is consumed to induce chemical changes in the resist, a finite attenuation is inevitable. The least exposure is at the bottom of the resist layer. Therefore, the developed positive resist profile in a homogeneous single resist layer has a larger opening on top, forming a so-called "overcut profile". The isotropic component in resist development also contributes to an overcut resist image profile just as in isotropic etching. With an overcut profile, a difference in dimension over topography results if the subsequent wafer delineation step, e.g., etching of SiO_2, is dependent on the resist dimension at the bottom (shown in Figure 3). The linewidth difference caused by the overcut profile is often referred to as the basic linewidth variation in the presence of underlying topography, without considering the optical interference effects described in the next paragraph. This can lead to significant errors.

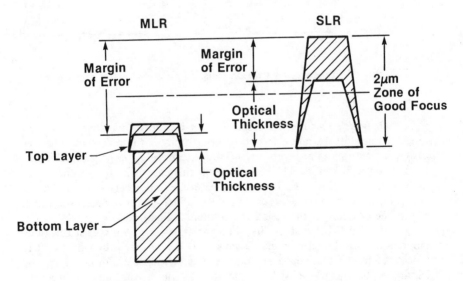

Figure 2. Comparison of focus tolerance of SLR and MLR systems.

6.2.a.3 Linewidth Variations due to Reflections off Topographic Features. A much greater cause of linewidth variation in optical lithography is due to optical interference effects caused by the imaging beam and its reflections off the surfaces of the wafer, topographic features and the resist (shown in Figure 4a). Beams 1 and 4 which undergo multiple reflections from horizontal interfaces produce standing waves in the resist image. The standing waves are periodic in the vertical direction, with a diminishing amplitude towards the top of the resist layer because optical attenuation produces a larger intensity difference in the interfering beams here. These standing waves account for linewidth variations even on planar substrates. Beam 2 is reflected off a vertical surface while beams 3 and 5 are scattered by edges. These beams cause interference to occur in the random directions as well as the vertical direction, resulting in significant horizontal linewidth variations (*1*) as shown in Figure 4. These variations are clearly caused by interference effects rather than simple resist thickness variation with overcut resist image profiles because linewidth change occurs at many locations near the topographic edges but not one-to-one to the edges.

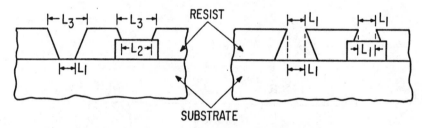

Figure 3.   *Linewidth variation caused by overcut resist profiles.*

6.2.a.4 Linewidth Variations Caused by Non-Uniform Reflectivity. A third cause of linewidth variation in optical lithography is the reflectivity changes caused by multiple film layers on the wafer, e.g., SiO_2 on polycrystalline Si on SiO_2 on a Si wafer. The reflectivity in this case is a strong function of individual layer thickness because of multiple interference. A thickness non-uniformity of any layer can produce a non-uniform reflection which affects the exposure of the resist, and ultimately linewidths.

With an MLR system, linewidth variations caused by resist thickness differences can be eliminated because of the planarization step. The reflections from either topographic structures or multiple film layers can be eliminated by using an absorbant material for the bottom layer. Either an inherently absorbant polymer material can be chosen or a dye can be added

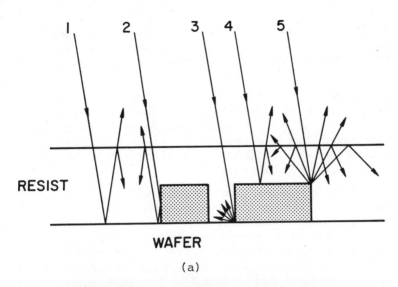

RESIST

WAFER

(a)

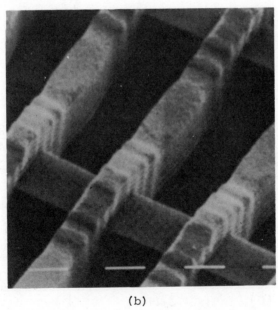

(b)

Figure 4. (a) Reflections from the substrate and topographic features. (b) SEM of 1.0-μm resist lines and spaces over a 0.5-μm SiO$_2$ step. The condenser-to-imaging lens aperture ratio σ indicating the degree of partial coherence is 0.5. (Reproduced with permission from Ref. 1.)

to the bottom polymer layer. However, if the refractive indices within the MLR structure are not unified, residual standing waves will still be present. These standing waves are now very uniform laterally and very low in contrast. They can easily be smoothed with post-exposure baking techniques (2).

6.2.a.5 Sensitivity.   Normally, when a thinner resist layer is used, the exposure time can be shortened as compared to that of a thicker identical material. However, the exposure time will be slightly longer in an MLR system because there is no component to exposure from the reflected light. Therefore, high sensitivity in an optical MLR system can be realized only by using a sensitive imaging resist.

6.2.b E-Beam Lithography.   With electron beam optics, a much lower NA is utilized to achieve submicrometer resolution and the depth of focus is much improved over optical lithography. However, once the electrons enter the resist and the wafer, electron scattering takes place, producing a cloud of forward and backward scattered electrons. Figure 5 shows simulated trajectories of 100 electrons scattered by a 0.4 μm poly(methyl methacrylate) (PMMA) film coated on a Si wafer (3). The calculated absorbed energy density, i.e., energy used to expose the resist, is shown in Figure 6 for forward and backward scattered contribution in a PMMA film (4). Therefore even though an electron beam can easily be focussed to less than 50 nm in

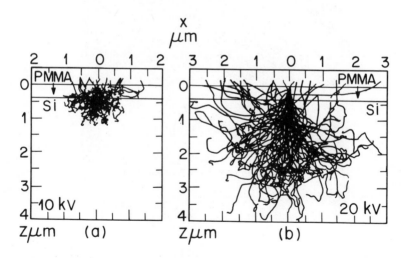

Figure 5.   Monte Carlo simulation of the trajectories of 100 electrons scattered in a 0.4-μm PMMA film coated on a Si substrate. (Reproduced with permission from Ref. 3.)

diameter, scattering produces an effective blurring of up to several micrometers in diameter. When exposed patterns are in the proximity of each other, these scattered electron clouds build up in unexposed areas to affect the exposure and reduce contrast, hence, the term "proximity effect." Another problem in e-beam lithography caused by the presence of a resist film is the "charging effect." When electrons incident on a thick resist layer, which is almost always is an insulating material, or with even a smaller resist thickness when the substrate is an insulating material, the electron charge can accumulate on the insulation surface and deflect the oncoming electrons. These two problems can be alleviated with an MLR system.

6.2.b.i Proximity Effects. Backscattering is usually more dominant than forward scattering because of a larger affected area, as exemplified by Figure 6. The backscattering coefficient is approximately proportional to the

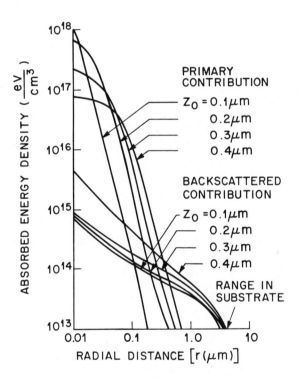

Figure 6. Calculated absorbed energy density due to forward and backward scattered electrons. Z_0 is the penetration depth in a 0.4-μm PMMA film which is coated on an Al substrate. The electron energy used is 20 keV. (Reproduced with permission from Ref. 4.)

atomic number of the scatterer (5). Therefore, backscattering from Si, atomic number 14 and Al, atomic number 13, is more significant than backscattering in an organic photoresist which consists of C, H, and O with atomic numbers 4, 1 and 8, respectively. In an SLR system, the resist suffers from electron backscattering from the wafer to the resist. With an MLR system, the imaging layer is separated from the wafer by another low-atomic-number organic material resulting in a reduction of the proximity effects. Figure 7 shows calculated radial distribution of the absorbed energy density at the midpoint of a 0.5 - μm thick resist, first in single layer form, secondly over 50 nm of Al and a 2 - μm thick planarizing resist layer, thirdly over only the 2 - μm planarizing layer, and finally over a 4 - μm planarizing layer on a Si wafer (6). The spread of the absorbed energy decreases as a function of the distance from the 0.5 - μm resist to the Si surface and to the Al surface as predicted. The Al layer, with an atomic number close to that of Si, produces only a slightly larger spreading from

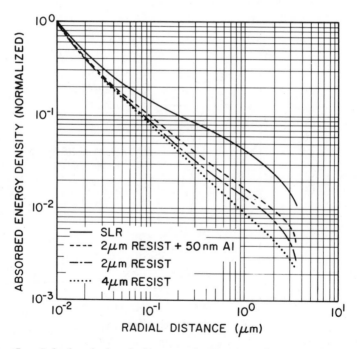

Figure 7. Calculated absorbed energy density as a function of radial distance for an infinitessimal beam. The energy density at the midpoint in a 0.5-μm imaging resist layer in SLR and various MLR configurations is plotted. (Reproduced with permission from Ref. 6.)

the corresponding MLR case without Al. Therefore, MLR enables better linewidth control, as shown in Figure 8a and 8b where the development of the exposed 0.5 - μm wide resist image in the 0.5 - μm thick imaging layer is simulated (6). The SLR case shown in Figure 8a clearly shows less image contrast and development tolerance than the MLR case shown in Figure 8b. Figure 9b shows a reduction of proximity effects as exemplified by the vertical resist profile in the top layer consisting of an identical material and thickness which is spun on a 2 - μm thick planarizing layer. For clarity of comparison, the planarizing layer was not delineated.

6.2.b.2 Charging Effects.   In order to prevent charging, a thin conducting layer, typically 50 nm of Al or Au, is coated on top of an SLR system. With an MLR system, a conductive imaging or middle layer can be used for the same purpose thus incorporating the extra coating step into a normal MLR process. There is also a possibility to heavily dope a semiconducting middle layer such as Si or Ge to prevent charging.

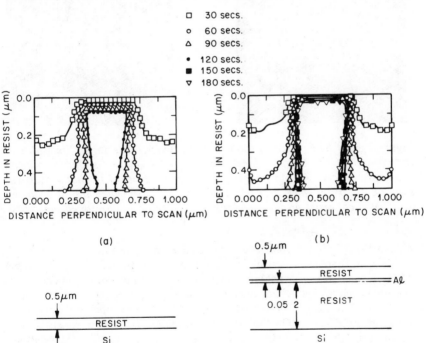

Figure 8.   (a) Resist images at different stages of development. 0.5 μm of PMMA was spun directly on Si. (b) Same as (a) but the resist is now spun on 0.05 μm of Al on another 2.0-μm of planarizing organic material. (Reproduced with permission from Ref. 6.)

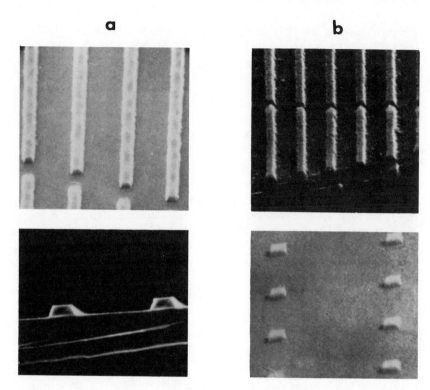

Figure 9. (a) SLR directly spun on a Si wafer. The larger linewidth at the bottom of the 0.5-μm thick negative resist images is due to e-beam proximity effects. (b) An imaging resist layer identical to that in (a) was spun on a 2.0-μm thick planarizing layer. The delineated top layer shows vertical resist profile, indicating reduction of proximity effects.

6.2.b.3 Sensitivity. In order to achieve an economically feasible throughput for e-beam systems, a sensitivity of the order of 1.0 $\mu C/cm^2$ is desirable. Many e-beam resists capable of high aspect ratio imaging cannot meet the sensitivity requirement. For example, PMMA requires about 80 $\mu C/cm^2$ in a 20-25 keV system. The sensitivity of PMMA-based copolymers are in the range of 30 $\mu C/cm^2$. On the other hand, many fast e-beam resists capable of meeting the 1 $\mu C/cm^2$ requirement such as poly(butene-1-sulfone) [PBS] (7), poly(glicidyl methoacrylate-co-ethylacrylate) [COP] (7), polyfluoroalkylmethacrylates [FBM] (8), and Sel N (9), cannot be used for high aspect ratio images. These fast resists can be candidates for the imaging layer of MLR systems which can transfer the image from the thin top layer to the thick bottom layer to exhibit aspect-ratio amplification.

6.2.c X-Ray Lithography. When a synchrotron storage ring is used as an x-ray source, an SLR system is adequate. The x-ray output from a storage ring is of sufficient intensity so that MLR is not required to amplify the aspect ratio. The beam is also sufficiently collimated to maintain a uniform resist profile within the exposure field. However, when a conventional e-beam-generated x-ray source is used, resist sensitivity becomes a major limiting factor of throughput and an MLR system is desirable. In addition, when the exposure field is comparable to the source-to-wafer distance, the x-ray beam becomes significantly oblique near the edge of the field, hence tilted resist image profiles. An MLR system can transform the tilted profile in the top imaging layer into vertical one in the bottom planarizing layer.

6.2.c.1 Sensitivity. The conventional x-ray imaging system operates at 3-20 kW in the 4 to 10 Å wavelength region. The available x-ray output is such, that in order to achieve an exposure time of about 1 minute for each wafer, a sensitivity in the order of 5 mJ/cm^2 is required. A similar resist sensitivity problem arises as in the e-beam case. Resists that can produce high aspect ratio images such as PMMA and FBM (*10*) requires high exposure dosages of 1 J/cm^2 and 50 mJ/cm^2 respectively. The two fastest resists, DCOPA (*11*)[poly(2,3-dichloro-1-propylacrylate) + poly(glycidyle methacrylate-co-ethyl acrylate)] and DCPA-NVC (*12*) [poly(2,3-dichloro-1-propyl acrylate)] + N-vinylcarbazole] require abut 10 mJ/cm^2. The usable thickness of these two resists is about 0.2 μm. Therefore, it is important to use these fast resists in an MLR configuration to satisfy the aspect ratio requirement.

6.2.c.2 Profile Restoration. With an x-ray point source, the resist image near the edge of the field is no longer vertical because of the oblique exposing beam, as shown in Figure 10a. This oblique resist image can affect the linewidth bias of critical features. Using a three-layer MLR system, a normal profile can be produced. Figure 10b shows typical procedures to restore the resist image profile. The middle isolation layer can be delineated with an isotropic etching technique so that the linewidth variation due to an oblique profile is minimized. The bottom planarizing layer can now be delineated with an anisotropic etching technique or a normal incident blanket exposure.

6.2.d Ion-Beam Lithography. Because ions are much heavier than electrons, ion scattering is not sufficient to produce proximity effects as in the case of e-beam exposure. Ions are also more efficient in exposing the resist material. Of course, ion beam lithography has its own problems. Before becoming a viable production technology one particular problem can easily be alleviated with MLR. Ions have a much shorter stopping distance in the resist and the substrate, typically in the order of nanometers depending on the type and energy of the ion. With light ions such as H or He, a 1 - μm

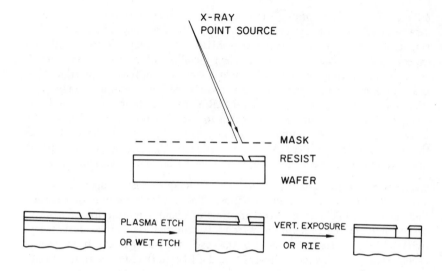

Figure 10. (a) *Oblique resist profile due to a large angle of incidence.* (b) *Vertical resist image profile obtained by blanket exposure or RIE of the bottom planarizing layer.*

penetration depth is possible, but with Ar or Ga ions only a fraction of a micrometer can be penetrated, as shown in Figure 11. Ion bombardment can also damage the semiconductor as is well known in the field of ion implantation technology. Therefore, when an MLR system is used, not only can the resist in the image layer be fully penetrated, but the bottom planarizing layer serves as a buffer to reduce ion bombardment damages to the semiconductor. Ions that are a good RIE barrier can be imbedded in the top portion of an SLR to form an exposure-induced RIE PCM, achieving MLR performance with SLR simplicity, as shown in Figure 12.

6.3 Existing MLR Systems

The evolution of MLR systems can be characterized by three periods: 1) the lift-off era, 2) the renaissance era, and 3) the improvement era. Originally, MLR systems were used to form a resist lift-off profile. The necessity of micrometer feature size, high aspect ratios, and linewidth control over topography were not felt until it was clear that Si-based memory would surpass magnetic bubble memory in performance and cost effectiveness. Most lithographers working on advancing the state of the art of lithography for bubbles realized that they had unique improvements for Si IC fabrication if

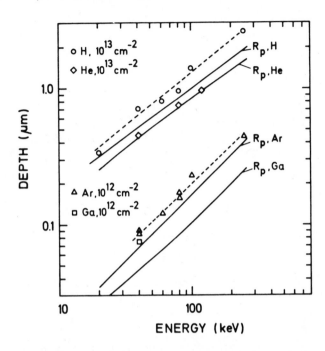

Figure 11. Projected ranges of Ar, H, He, and Ga and the developed thickness removed of Dupont Elvacite 2008 PMMA after exposure with 10^{13} cm^{-2} H and He, 10^{12} cm^{-2} Ar and Ga, respectively. (Reproduced with permission from Ref. 13.)

they could control the feature size over severe topographies. MLR systems were then revived for planarization. Gradually the advantage of MLR systems listed in section 6.2 were recognized for all branches of lithography. In the last two years there has been an exploding growth of literature on MLR systems and applications as MLR entered into the improvement era. After covering the three eras, we will discuss some modified single layer systems. These systems exhibit some MLR effects, yet are simpler to process. Though their performance has not yet reached that of MLR systems, they offer a reasonable compromise as well as a suggestion of the direction for future MLR systems.

6.3.a The Lift-Off Era of MLR.   In 1973, the minimum feature size of manufactured IC's was of the order of several micrometers; resolution, aspect ratio, and the linewidth control requirements could easily be satisfied with SLR. However, there was a need to produce an inverted resist profile,

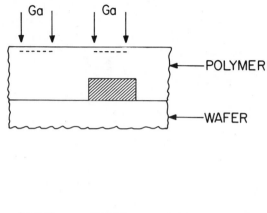

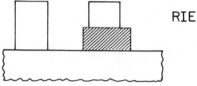

*Figure 12.   Ga ion implanted into a thick resist layer to form an in-situ
RIE PCM.*

i.e., openings narrower on top than at the bottom of the resist image, so that
when a metal is anisotropically deposited, a discontinuity takes place
between the unwanted metal on the resist and the desired metal pattern in
the resist opening. Submerging this wafer with the inverted profile in a solvent, lifts off the unwanted metal with the resist, hence the term lift-off.

Havas et al (14) used a three layer structure to produce the inverted
profile. The bottom planarizing layer was either a cross-linked negative
resist such as KTFR or a hard baked positive resist such as AZ111. An
inorganic material, preferably metal, was used as the middle layer, as well
as the RIE PCM. The top layer could be any conventional resist. The process is straightforward by modern MLR standards. After the top imaging
layer is delineated, the inorganic layer is wet etched. Then an O_2 plasma
etch or RIE delineates the planarizing layer with and without an inorganic
overhang, respectively.

Evaporated Al was used by Greber et al (15) as a near UV PCM,
followed by Chamberlin and Bergeron (16) in 1974 who used evaporated Si.
Subsequently, Havas (17,18) overcame one of the major process complications of the three - layer system, namely a vacuum deposition of the middle

inorganic layer. Switching to polydimethysiloxane as the RIE PCM, he could use the spincoating technique. The transparent property of the siloxane material also eliminated the inability to see through a metallic middle layer for alignment. The so-called "spin-on-resin glass" was an Owens-Illinois type 650 resin dissolved in *n*-butyl acetate. A thickness of about 150 nm was used. This siloxane material was referred to as organosilicon in the 1976 article of Havas (*18*).

Figure 13a shows the resist in a three-layer Al-RIE-PCM system by Havas et al. The dry etching conditions were similar to those of plasma etching. Therefore, an undercut was produced under the Al PCM to facilitate lift-off. Figure 13b shows the resist image in a three-layer siloxane RIE PCM system. The dry etching process was sufficiently anisotropic to eliminate the undercut.

Subsequently, other RIE PCM materials were used. They were selected either for their RIE characteristics or the feasibility of the deposition process. A list of these systems is shown in Table I. During this period, Hatzakis (*19*) developed a two-layer wet etch PCM system for e-beam exposure. Basically, a PMMA copolymer is spun on PMMA, or vice

 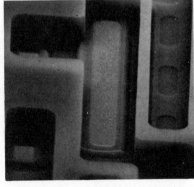

a   b

Figure 13.   (a) Resist images obtained by using Al as the RIE PCM in a three-layer system. The isotropic component produced an undercut beneath the Al PCM for lift-off. (b) Resist images obtained by using spin-on polymethylsiloxane as the RIE PCM in a three-layer system. There is no undercut in this case. (Reproduced with permission from Ref. 18.)

TABLE I. MLR Systems in the Lift-Off Era

Masking Material	Function	Layers	Authors	Year
Inorganics	RIE-PCM	3	Havas et al (14)	1973
Al	Near-UV PCM	3	Greber et al (15)	1974
Si	Near-UV PCM	3	Chamberlin (16) & Bergeron	1974
Siloxane	RIE PCM	3	Havas et al (17)	1975
Si	RIE PCM	3	Zielinski et al (20)	1976
Copolymer	Wet-Etch PCM	2	Hatzakis (19)	1977
SiO	RIE PCM	3	Zielinski et al (22)	1977
Si_3N_4	RIE PCM	3	Havas et al (18)	1978
MgO	RIE PCM	3	Zielinski et al (21)	1979

versa. Hatzakis identified mutually exclusive developers for the two layers so that the development of the top layer dissolves the bottom layer very slowly, with a dissolution rate merely sufficient to break through the interfacial layer of the two resists. This two-layer system produces the best profile when the slower resist is coated on the faster resist. However, in order to attain an acceptable e-beam sensitivity of 30 $\mu mC/cm^2$, the faster copolymer resist is often used for the top imaging layer, leaving the bottom PMMA layer barely exposed. Therefore, the development of the bottom layer is very isotropic, assimilating a wet etch situation, hence the term "wet-etch PCM." This MLR system is simple to use. In addition to reducing the number of layers to two, it requires only one exposure and two developments. Figure 14 shows the resist image profile of the copolymer/PMMA two-layer wet etch PCM system.

6.3.b The Renaissance Era. Near the end of the lift-off era, a new group of lithographers, aware of the difficulties of submicrometer linewidth control over topography, started using MLR systems to demonstrate the advantages discussed in Section 6.2. The lithography development community was awakened by the excellent results demonstrated. However, the manufacturing community was skeptical about MLR because of process complexity and

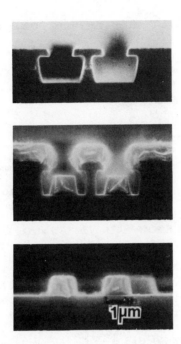

Figure 14.   Resist images of the two-layer copolymer wet-etch PCM system.   A 0.4-μm thick copolymer layer was spun on 1.4 μm of PMMA. (Reproduced with permission from Ref. 19.)

lack of a pressing need to produce micrometer and submicrometer devices. Major innovations in this period include the two- and three-layer deep-uv PCM systems, the two-layer inorganic RIE PCM system, the two-layer spun-on RIE PCM system, and the three-layer SiO_2 RIE PCM system, as shown in Table II.

6.3.b.1 The Two-Layer Deep-UV PCM System. Lin (*23*) discovered that certain diazo sensitized novolak - based resists such as AZ1350 exhibit strong absorption in deep-UV and demonstrated that it could be used as a deep-UV mask absorber, either in the conventional way of building the mask absorber on quartz substrates or by spinning the absorber directly on a deep-UV resist layer to form a so-called "Portable Comformable Mask." The two-layer deep-UV PCM system for near-UV and e-beam imaging was subsequently fully described in separate articles (*24-26*). Basically, the AZ imaging layer was first exposed with e-beam or near-UV light. After

TABLE II. MLR Systems in the Renaissance Era

Masking Material	Function	Layers	Authors	Year
Novolak resist	Deep-UV PCM	2	Lin (23)	1978
Ge-Se	RIE PCM	2	Tai et al (27,28)	1979
Al	Deep-UV PCM	3	Lin & Chang (25)	1979
SiO$_2$	RIE PCM	3	Moran (34) & Maydan	1979
Siloxane	RIE PCM	2	Hatzakis (31)	1981

development of the AZ layer, the wafer is subjected to a deep-UV blanket exposure. The bottom planarizing layer could then be developed with the AZ cap retained or removed, resulting in a capped or uncapped image, respectively.

The schematic processing steps for the two-layer deep-UV PCM system are shown in Table III for near-UV and e-beam exposures, resulting in capped and uncapped profiles respectively. Step 5a is required only for near-uv exposures to average out optical interference effects discussed in Section 6.2.a. Step 6a is used to enhance the AZ to PMMA adhesion and cap retention. Figure 15 shows uncapped 1 - μm lines in a 2 - μm thick PMMA layer. Figure 16 shows the capped image with 0.3 μm AZ on 2 - μm PMMA. The AZ layer was delineated by a 30 - μC/cm^2 25 - keV e-beam in both cases.

6.3.b.2 The Two-Layer Inorganic RIE PCM System. Tai (27,28) combined the silver sensitized chalcogenide glass material (29) Ge$_x$Se$_{1-x}$ with a planarizing organic layer to form an RIE PCM system. He also identified the "edge sharpening" effect as the cause of the unusual resolution and linewidth control characteristics of this two-layer system. After deposition of a 0.2 - μm Ge-Se layer on the organic planarizing layer by evaporation or sputtering, Ag is either directly deposited on the Ge-Se layer by evaporation or is deposited in the form of a compound such as AgCl or Ag$_2$Se. A wet deposition process using a bath containing Ag(CN)$_2$ anions is preferred (30). The Ge-Se at the bath interface undergoes the following reaction to become Ag$_2$Se.

$$GeSe_2 + 4KAg[CN]_2 + 8KOH \rightarrow 2Ag_2Se + K_4GeO_4 + 8KCN + 4H_2O$$

TABLE III. Processing Steps for the Two-Layer Deep-UV PCM Systems

		Uncapped	Uncapped	Capped	Capped
1.	Spin on PMMA	*	*	*	*
2.	Bake	*	*	*	*
3.	Spin on AZ1350	*	*	*	*
4.	Bake	*	*	*	*
5.	Expose AZ	*	*	*	*
5a.	Bake	*	*	*	*
6.	Develop AZ	*	*	*	*
6a.	Bake			*	*
6b.	Remove interfacial layer with O_2 plasma			*	*
7.	Deep-UV blanket exposure	*	*	*	*
8.	Dev. PMMA in toluene			*	*
8a.	Dev. PMMA in MIBK	*	*		

During the image exposure, Ag is photodoped into the Ge-Se layer. The Ag_2Se in the unexposed area is removed by a KI/KI_3 solution that converts the Ag_2Se to a soluble complex $KAgI_2$. Then the Ge-Se in the unexposed area is removed in a developer containing NaOH and a small amount of S^- anions. Now the underlying organic layer can readily be RIEed with O_2 using the photodoped Ge-Se as the RIE PCM. Schematic processing steps are outlined in Table IV.

The most important characteristics of this two-layer system is the so-called "edge sharpening" effect. During exposure, as Ag is doped into the Ge-Se layer, two other phenomena happen. The optical transmission of the Ag_2Se film increases as Ag is depleted. This enhances the exposure in the exposed areas just as the case of photobleaching of conventional organic

TABLE IV. Processing Steps for the
Two-Layer Ge-Se RIE PCM System

1. Spin on organic planarizing layer.
2. Bake.
3. Deposit Ge-Se.
4. Deposit Ag_2Se.
5. Expose.
6. Remove Ag_2Se from unexposed areas.
7. Develop Ge-Se in the nonphotodoped areas.
8. RIE planarizing layer.

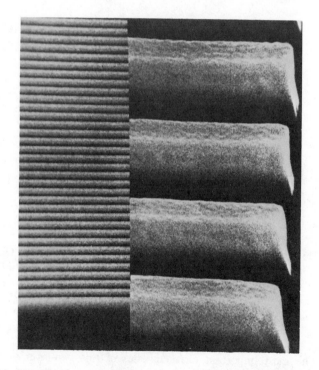

*Figure 15.   Uncapped resist image obtained with a two-layer deep-UV
PCM system.   The 0.85-μm wide PMMA lines were 1.9-μm thick separated
by 2.4 μm.   A 0.2-μm AZ1350J PCM was used.*

Figure 16.   Capped resist image obtained with an identical two-layer deep-UV PCM system as in Figure 13. The AZ layer was 0.3-μm thick on 2-μm thick PMMA.

resists. In addition, because of the gradient of Ag concentration at the image edges, Ag in the Ag_2Se layer is diffused laterally into the exposed regions. Even though the extra amount of Ag can reduce UV transmission somewhat, the lateral diffusion dominates to produce a higher concentration of doped silver in the Ge-Se layer image edges. Even though in principle Ag in a narrow unexposed strip can be exhausted and produce no more edge sharpening, the edge sharpening effect obviously is valid for features as small as 0.5 μm. Half-micrometer isolated lines, spaces, and gratings were printed with one single exposure dosage using a 0.31 NA, lens. Because of high opacity of the Ag_2Se/Ge-Se layers in the UV, optical interference effects from the underlying topography is isolated. Better yet, a high transmissivity at wavelengths above 500 nm enables alignment mask viewing. Figure 17 shows the combined Ge-Se/organic resist images after O_2 RIE. A thickness of 2.5 μm was used for the organic planarizing layer.

6.3.b.3 The Two-Layer Spun-On RIE PCM System. Hatzakis (*31*) combined polysiloxane (*32,32*) with a conventional planarizing layer to form a two-layer RIE PCM system which has a high sensitivity for e-beam and deep-UV exposures. Using either poly(vinyl methylsiloxane) [P(VMS)] or poly(dimethyl siloxane) [P(DMS)], an e-beam sensitivity of 2 $\mu C/cm^2$ and a contrast of 2 was reported. Deep-UV exposure with 220 - nm light also produced impressive results. Similar to any RIE PCM system, the RIE PCM layer can be coated on a hard-baked resist layer, such as AZ1350

*Figure 17.   Resist images delineated with a GCA DSW 4800 mask aligner.
The Ag$_2$Se/Ge$_{0.1}$Se$_{0.9}$/HPR two-layer inorganic RIE PCM system was used.
The HPR planarizing layer was 2.5 μm in thickness. ((Reproduced with
permission from Ref. 27.)*

baked to 200°C. The unique properties of P(VMS) and P(DMS) are that
they can be dissolved in a solvent and spin coated. These resists have to be
exposed without prebake. After exposure, the resist in the exposed areas is
cross-linked and will remain on the planarizing layer after a solvent development. An O$_2$ RIE can anisotropically delineate the bottom planarizing layer
to form a high-aspect-ratio image. Schematic processing steps of this two-
layer spun-on RIE PCM system is shown in Table V.

TABLE V. Processing Steps for the
Two-Layer Spun on RIE PCM System

1.   Spin on organic planarizing layer.

2.   Hard bake.

3.   Spin on siloxane.

4.   E-beam or deep-UV exposure.

5.   Develop.

6.   RIE planarizing layer.

The e-beam delineated 0.3 - μm thick P(VMS) image on the RIE's 3 - μm thick AZ1350J planarizing layer is shown in Figure 18a. A dosage of 2 - μC/cm^2 was used. A similar 0.3 - μm thick P(VMS) image on 2.5 μm of AZ1350J shown in Figure 18b was delineated with 220 nm deep-UV light with a very respectable sensitivity.

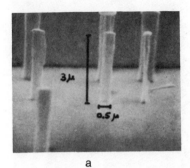
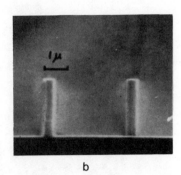

a   b

Figure 18.   (a) Submicrometer features delineated in a 0.3-μm thick P(VMS) layer on 3.0 μm of AZ1350J. The e-beam dosage used was 2.0 μC/cm^2. (b) Similar P(VMS) features on 2.5 μm of AZ1350J. The imaging layer was delineated with 220-nm deep-UV light. (Reproduced with permission from Ref. 19.)

6.3.b.4 The Three-layer RIE PCM System.   The MLR system is extremely similar to those used in the lift-off era. In fact any of those systems is applicable for modern high resolution and high aspect ratio imaging. Moran and Maydan (*34*) were first to demonstrate the versatility and process compatibility of three-layer SiO$_2$ RIE PCM system. E-beam, x-ray, refractive optical projection, and reflective optical projection were used by them to delineate the top imaging layer. Patterning of silicon nitride, polysilicon, phosphosilicate glass, and aluminum using this three-layer RIE PCM system was claimed by them. The schematic processing step is shown in Table VI. Figure 19 shows 1 - μm wide SiO$_2$ lines on the RIE'd planarizing layer in a three-layer SiO$_2$ RIE PCM system.

The three-layer deep-UV PCM system of Lin and Chang (*34*) used Al as a deep-UV PCM instead of SiO$_2$ as the RIE PCM. This eliminates the need for an RIE system but still requires a large number of processing steps. Figure 20 shows a 0.3 - μm thick Al deep-UV PCM on 1.6 μm of PMMA.

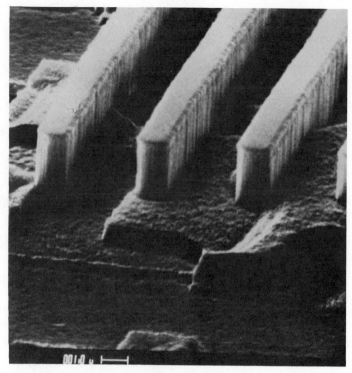

Figure 19. SiO₂ lines on the RIE'd planarizing layer crosses over topographic features on the wafer. The SiO₂ three-layer RIE PCM was used to delineate these 1.0-μm lines. (Reproduced with permission from Ref. 34.)

TABLE VI. Processing Steps of the
Three-Layer SiO_2 RIE PCM System

1. Spin on planarizing layer.
2. Hard bake.
3. Plasma CVD or sputter coat SiO_2.
4. Spin on imaging layer.
5. Bake.
6. Expose.
7. Bake.
8. Develop imaging layer.
9. Descum.
10. Etch SiO_2.
11. RIE planarizing layer.

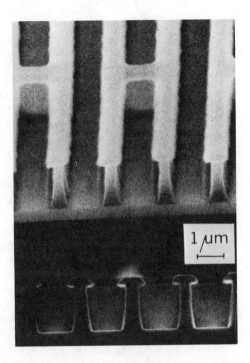

Figure 20.   A 0.3-μm thick Al deep-UV PCM on 1.6 μm of PMMA. The pattern has a periodicity of 2.0 μm.

TABLE VII. Processing Steps for the
Three-Layer Deep-UV PCM System

1. Spin on planarizing layer.
2. Bake.
3. Deposit Al.
4. Spin on imaging layer.
5. Bake.
6. Expose.
7. Bake.
8. Develop imaging layer.
9. Descum
10. Etch Al.
11. Expose planarizing layer.
12. Develop planarizing layer.

6.3.c The Improvement Era. The improvement era started in 1980 following major discoveries between 1978 and 1979. The siloxane system of Hatzakis (*1*), though reported in 1981, is a unique and promising system. Therefore it was included in the renaissance era. The MLR systems have evolved into either RIE PCM or deep-UV PCM by this time and many of the researchers in this period have been more concerned with applying the MLR principle to improve lithographic performance. A list of the MLR systems reported in this period is shown in Table VIII. Because of the increased volume of publication in this period, the list is meant to be representative rather than exhaustive.

Silicon has been a popular choice for the middle layer in three-layer systems because of many reasons. 1) It is a routinely-used material in Si technology, with known deposition and etch characteristics, and inherently does not pose a contamination or safety problem. 2) It is a good O_2 RIE etch barrier. 3) It has favorable spectral characteristics, namely opacity in deep-UV for delineation of the planarizing layer and transmission in visible light to facilitate alignment. The spectral transmission of Si is shown in Figure 21. 4) It can be doped to enhance conductivity and thus reduce e-beam charging effects. Ge (*37*) and R_i (*45*) have also been in three-layer RIE PCM systems and may be considered to be subsets of the Si group. A more than usual stress problem between Ti and resist was encountered.

The three-layer SiO_2 RIE PCM system received several refinements. O'Toole et al (*38*) included a dye in the planarizing layer, so that the latter not only planarizes but also absorbs the incoming and reflected light beams thereby eliminating the optical interference effects discussed in Section 6.2.a. The improvement is clearly seen in Figure 22, where the resist image over a 0.5 - μm step is shown in the SLR, undyed MLR, and dyed MLR forms, respectively. Another significant improvement was made by replacing dry deposition of SiO_2 with a spin-on process (*39,40*) which is virtually a reinvention of the Havas process (*17*).

The two-layer novolak deep-UV PCM system also received several refinements. Batchelder and Takemoto (*43*) and Griffing et al (*44*) succeeded in building a low pressure Cd arc lamp deep-UV source that is highly efficient in deep-UV and possesses the proper spectral cut-off for the PMMA planarizing layer. Existing Xe-Hg arc lamp deep-UV sources produce less than 1% of the input electric power in the region of 200-250 nm. It also has a dominant output around the 254 - nm Hg line which reduces the contrast of the novolak mask absorber because there is a slight transmission window in that spectral region for novolaks as shown in Figure 23. The Cd lamp can convert 10% of the 100-watt input power into deep UV in the spectral range of 200-230 nm without the 254 nm peak. Therefore, its spectrum is very desirable for the novolak deep-UV PCM technique. Another promising deep-UV blanket exposure source is the microwave-powered elec-

Table VIII

MLR Systems in the Improvement Era

Masking Material	Function	Layers	Authors	Year
Si	RIE PCM	3	Watts et al (*35*)	1980
			Kruger et al (*36*)	1981
Ge	RIE PCM	3	Tennant et al (*37*)	1981
Si_3N_4	RIE PCM	3	O'Toole (*38*)	1981
Ge-Se	Deep-UV PCM	2	Ong et al (*41*)	1981
Novolak Resist	Deep-UV PCM	2	Batchelder (*43*) & Takemoto	1981
			Griffing et al (*44*)	1981
			Ong et al (*41*)	1981
			Santini (*46*) & Viswanathan	1982
PSTTF	Deep-UV PCM	2	Lin et al (*47*)	1982
Ge-Se	RIE, Deep-UV PCM	3	Vadimsky et al (*42*)	1982
Si	Deep-UV PCM	3	Bassous et al (*48*)	1982
SiO_2	Rie PCM	3	Ray (*39*) Ting et al (*40*)	1982
Ti	RIE PCM	3	Bassous et al (*48*)	1982

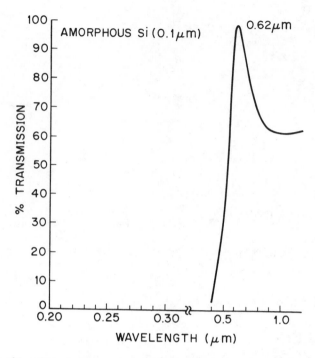

Figure 21. Spectral transmission of a 100-nm layer of hydrogenated amorphous Si deposited by plasma CVD of a silane/helium mixture on a Suprasil quartz substrate. (Reproduced with permission from Ref. 53.)

trodeless Hg lamp (*43*). Microwave excitation of Hg produces more deep-UV output than conventional Hg arc lamps. An efficiency of better than 8% in the region 200-260 nm has been claimed. Provided the input power in the kW range, the output deep-UV power can be in the order of 100 watts. Currently, only about 15% of this deep-UV power can be collected because of the special geometric restrictions imposed by the microwave cavity. In addition to the immense deep-UV output, the electrodeless lamp is potentially more stable than the Hg or Cd arc lamps. It has been reported that there is only a 10% reduction in output after 500 hours of operation was claimed. Unlike arc lamps which require long warm up time, these electrodeless lamps need only 3 seconds to reach full power and therefore can be turned off completely between exposures. Santini and Viswanathan (*46*) used a plasma treatment step with the PMMA layer before application of the novolak layer to prevent formation of the novolak-PMMA interfacial layer. They demonstrated 0.8 - μm resolution from a Mann 4800 aligner using a 0.28-NA 436 - nm lens as shown in Figure 24.

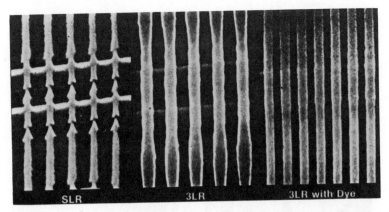

Figure 22.   (a) SLR image of 1.0-μm lines and spaces over 0.5 μm of polysilicon steps.   The resist thickness was 1.0 μm.   A lens of 0.28-NA and 436-nm wavelength with a coherence factor σ = 0.7 was used.   (b) 1.0-μm lines and spaces of SiO₂ three-layer RIE PCM on 1.5 μm of Hunt 204 resist over a similar step using an identical lens.   (c) Similar to (b) except that the bottom planarizing layer was dyed.   (Reproduced with permission from Ref. 38.)

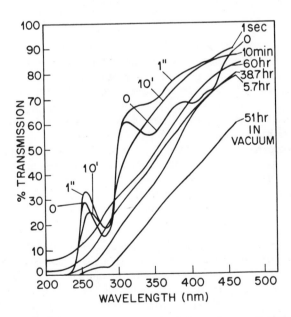

Figure 23.   Spectral transmission of a 0.2-μm layer of AZ1350J PCM.

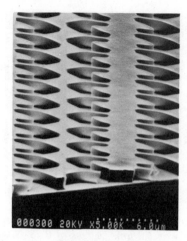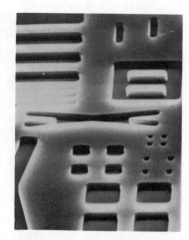

Figure 24. Uncapped images delineated with the Novolak/PMMA deep-UV PCM system. A GCA Mann DSW4800 aligner with 0.28-NA and 436-nm wavelength was used. The smallest openings are of 0.8 μm nominal dimension. (Reproduced with permission from Ref. 46.)

The PSTTF(polystyrene-tetrathiafulvalene) deep-UV PCM system (*47*) demonstrated that a negative e-beam resist can be used as the top layer of a deep-UV PCM system. In principle, when a negative resist is used for the top layer, the deep-UV resist is exposed by e-beam in the undesirable areas. However, if there is a sufficient sensitivity ratio between the top and the bottom resists, as in the case of PSTTF which requires only 10 $\mu C/cm^2$ to expose compared to 80 $\mu C/cm^2$ for PMMA, the e-beam exposure of the bottom layer can be tolerated. PSTTF exhibited negligible interfacial mixing with PMMA. A 0.3 - μm thickness was reported to be an excellent deep-UV mask for the underlying PMMA layer and enabled high resolution, high aspect ratio patterns to be fabricated as shown in Figure 25.

6.3.d Single Layer Alternatives. Single layer systems can be modified or designed to simulate some multilayer effects and thereby produce better imaging characteristics. Two particular systems will be discussed here. The first system utilizes a soak to reduce the dissolution rate of the resist near the top surface. The second system relies on favorable spectral absorption and spectral sensitivity characteristics to produce some decoupling of the surface and the bulk behavior of the resist.

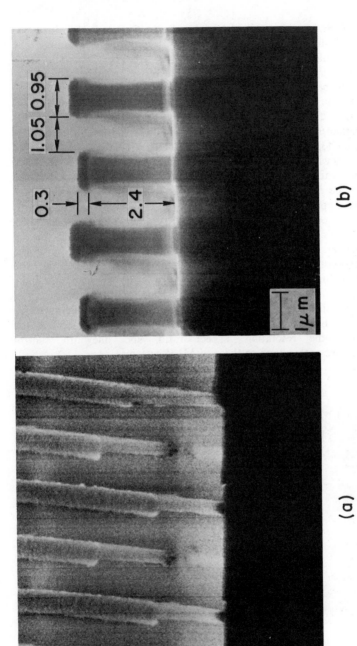

(b)

(a)

Figure 25.   Nominally 0.5 μm (a) and 1.0 μm (b) PSTTF lines spaced 1.0 μm apart in 2.4-μm thick PMMA (Elvacite 2041). The PSTTF caps are 0.3-μm thick delineated with a 10-μC/cm² e-beam dosage.

6.3.d.1 The Chlorobenzene-Soaked Novolak Resist System. Hatzakis et al
(*49*) subjected wafers with diazo-sensitized novolak resist to a chlorobenzene
soak prior to development, either before or after exposure. The chlorobenzene penetrates approximately 300 nm into the resist, reducing the dissolution rate of the penetrated layer. This reduction is sufficient to overcome
the natural overcut resist image profile caused by both the optical absorption
characteristics and the isotropic nature of the resist development, as discussed in Section 6.2.a. The resultant soaked resist image profile has a narrow opening on top, as shown in Figure 26. This profile can now be used
for lift-off. Linewidth control is better than that of untreated single layer
resist images because 1) the dimension controlling layer is at the top of the
resist and 2) the reflected imaging beam that causes multiple interference
effects is least appreciable at the top of the resist since the optical absorption effect produces the largest contrast in the standing waves. Besides lift-
off, anisotropic etching can also benefit from this soaked profile, if the top
300 nm thickness adequately masks the etching. Table IX shows the
schematic processing steps of this soaked system. As mentioned earlier, the
soak can be performed before or after exposure. The post-exposure bake
enhances the surface effect and smoothes the standing wave image. The
performance of this system is clearly better than conventional one-layer systems. However, because of incomplete decoupling of the surface and the
bulk behavior, the resolution, aspect ratio, and linewidth control
characteristics are not as good as those of MLR systems.

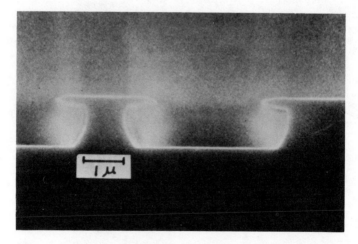

*Figure 26.   A one-layer novolak-based resist image after chlorobenzene
soaking.*

TABLE IX. Schematic Processing Steps of the
Chlorobenzene-Soaked Novolak System

1. Spin on the novolak resist.
2. Prebake.
3. Soak in chlorobenzene (may be done after Step 4).
4. Expose.
5. Bake.
6. Develop.

6.3.d.2 The Spectral Selection System. The negative deep-UV resist MRS (*50*) responds to deep-UV and mid-UV light. Its deep-UV sensitivity is higher than that in the mid-UV. In addition, absorption in the deep UV is greater than in the mid UV. This leads to a high-resolution surface delineation because of the dominating effect of the shorter wavelengths at the surface. The deep penetration of the longer wavelengths helps to support a high aspect ratio. Linewidth control is very good because of the narrower resist opening at the top, just as in the soaked system. This unusual spectral behavior is demonstrated in Figures 27a and 27b (*51*). The former shows the resist image delineated by 222-nm light produced by a KrCl excimer laser. The resist line at the top if extremely wide and narrows quickly because off insufficient light penetration into the bulk of this negative resist. On the other hand, when 308 nm light from a XeCl excimer laser was used, an almost vertical resist profile was produced, showing very good light penetration with the longer wavelength. Again, because of incomplete decoupling of surface and bulk characteristics, the performance of this resist system is not as good as true MLR systems, but is better than untreated SLR system. Table X shows a schematic process sequence for the MRS resist.

6.4 Practical Considerations

6.4.a Planarization. As discussed in Section 6.2.a, a thickness difference in the resist film leads to linewidth variations. This also applies to the imaging layer in an MLR system. When the surface of the "planarizing" bottom layer is not perfectly planar, the surface of the crucial top imaging layer is not planar either and the resultant thickness variations will lead to linewidth changes. In principle, perfect planarization by the bottom layer is impossible. The most obvious situation is in the case of a single semi-infinite step on a wafer as shown in Figure 28. The resist at the extreme sides of the step has to be identical in thickness, (equal to a in Figure 28) leaving the area in the vicinity of the step impossible to planarize. Therefore, in reality,

a

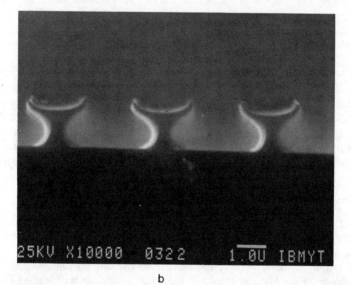

b

Figure 27. (a) *Negative image in Hitachi RD2000 exposed with a CrCl excimer laser at 222 nm. The extremely small dimension at the bottom indicates heavy light absorption.* (b) *Same as in* (a) *except XeCl laser-producing 308-nm light was used.* (*The absorbance is less at this wavelength.*)

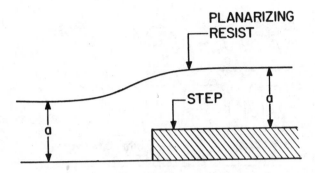

Figure 28. A semi-infinite step that is theoretically impossible to planarize.

TABLE X. Schematic Processing Sequence of MRS Resist

1. Spin coat resist.

2. Prebake.

3. Expose.

4. Develop. Rinse sequentially in two different solvent baths before deionized water rinse.

the function of the planarization layer is to smooth the undulation caused by topographic features on the wafer so that the top imaging layer can have a much smaller thickness variation.

Rothman (52) investigated the planarization of polyimide films over features tens of micrometers in size and separation. Bassous and Pepper (53) studied planarization of PMMA and AZ1350J over features pertinent to Si wafer processing. A mechanical stylus was used to determine the topography of the wafer and the corresponding surface variation of the resist thickness as shown in Figure 29 where a 1.7 - μm thick AZ1350J layer was spun on steps of different space and width combinations.

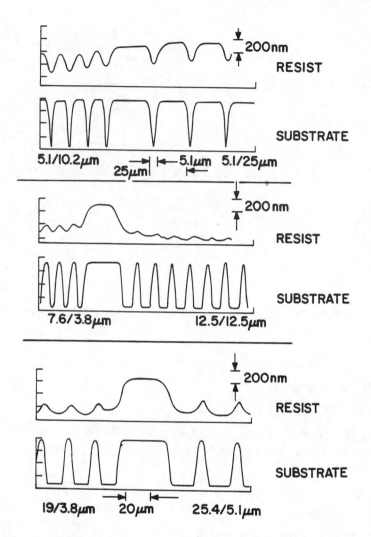

Figure 29. The surface profile of a nominally 1.7-μm thick AZ1350J shown with the corresponding underlying topography. (Reproduced with permission from Ref. 53.)

The effect of molecular weight of the resist on degree of planarization is shown in Figure 30a where the amplitude of fluctuation of 2 - μm thick PMMA films with average molecular weight of 500,000, 150,000, and 33,000 respectively, is plotted as a function of periodicity of 0.8 - μm high and 5 - μm wide steps. Resists with a lower molecular weight clearly planarize better. However, the contrast of a positive resist such as PMMA usually increases with molecular weight and should not be arbitrarily reduced just to improve planarization in MLR application. In the case of RIE PCM, the other properties related to molecular weight such as temperature stability and RIE rate should be balanced against the planarization requirement.

The planarization effects of *multiple* coating are illustrated in Figure 30b where PMMA of an average molecular weight of 500,000 was used for 1, 2, 3, and 5 coatings with a 160°C 1 hour bake in between. While a thicker film planarizes better, a film of a smaller absolute thickness can planarize better than a thicker film if the number of applications is increased, as exemplified by the comparison of a 2.5 - μm thick coating obtained with five 0.5 - μm applications, with a 3 - μm coating and with two 1 - μm applications. Another example is seen from a comparison of the curve corresponding to a 500,000 average molecular weight in Figure 30a with that in Figure 30b for two 1 - μm coatings. This multiple planarizing technique has no effect on the resist imaging, etching, or temperature properties. The obvious drawback to this technique is that processing becomes more complex.

Planarization is also a function of the width and space of the topographic features as shown in Figure 30c where a 2 - μm thick PMMA film with average molecular weight of 500,000 was spun on 0.8 - μm high topographies of the indicated width and space at varying periodicity. Obviously, large features and spaces are more difficult to planarize. Fortunately, the planarization requirement on large features usually is less critical than that on small features.

6.4.b Pinholes.   Because a *thin* top layer is used to produce better resist images, MLR is more susceptible to pinhole problems. Therefore, the pinhole density of each MLR system has to be evaluated to qualify for production runs. As it is tedious and unreliable to actually count pinholes, an MOS capacitance test developed by Fatula (*54*) to compare pinhole density in resist films can be modified to test MLR systems. The entire MLR layers can be spun on a 35 - nm thick SiO_2 film, then processed with all the pertinent processing steps for the particular MLR system except for the exposure of the imaging layer. The SiO_2 film with the MLR stack still attached is now subjected to a BHF etch or CF_4 RIE to facilitate removal of SiO_2 at the pinhole sites. The resist is then stripped and 0.8 - mm diameter Al dots can be deposited on the SiO_2. The total number of shorts when the applied voltage to the Al dots is at 1.5 and 3.0 mV/cm is divided by the total area of the Al dots to indicate the pinhole density. This test can be modified further to identify the contribution of the pinholes from a specific

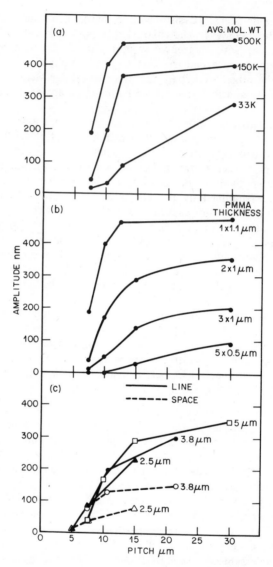

Figure 30. (a) *Amplitude of the surface variation of a 2.0-μm thick PMMA film covering 0.8-μm high topography. The steps are 5.0-μm wide and the spaces are varied. Three types of PMMA with average molecular weights, 33,000, 150,000, and 5000,000 were used.* (b) *Same as* (a) *using only the PMMA with an average molecular weight of 5000,000. One application of a 1.1-μm film, accumulation of two layers of 1.0-μm film, three layers of the same, and five layers of 0.5-film are compared.* (c) *Same as* (b) *with a single application of a 2.0-μm thick PMMA film. Step widths of 2.0, 3.8, and 5.0 μm with varying spaces and spaces widths of 2.5 and 3.8 μm with varying step widths are compared.*

processing step. This MOS pinhole test is functional and objective. However pinhole size variation and multiple pinholes under each Al dot cannot be detected.

Such a pinhole density test was performed on the AZ/PMMA two-layer deep-UV PCM system (*26*). The result is shown in Table IX where a pinhole density of 8 and 6 per cm^2 was obtained for the capped (A) and uncapped (B) systems. Because only three wafers were used for each test, the result should be taken only qualitatively and the numerical difference between 6 and 8 pinholes/cm^2 should be taken as being indicative of measurement fluctuations only. It should not be attributed to the use of different developers or O$_2$ plasma because in the subsequent tests of batches C and D in which the DUV exposure was omitted, the numbers were 0 and 1 pinhole/cm^2 with the *capped* system giving the smaller pinhole density. The low pinhole density in batch E in which the AZ *development* step was omitted suggests that the pinholes arise during the development of the AZ layer. Presumably, a small portion of the AZ base resin molecules were not linked up with the photoactive compound and therefore still exhibited their intrinsic high solubility in the AZ developer. After development, these high solubility spots became pinholes. These pinholes are apparently larger than the diffraction - limited sizes so that they can be transferred into the PMMA film by deep-UV exposure.

TABLE XI. Pinhole Test Results of the
AZ/PMMA Two-Layer Deep-UV PCM System

Batch	AZ dev.	O$_2$ plasma	Deep-UV exposure	MIBK dev.	ϕ-Cl dev.	Pinholes per cm*ref*2
A, full capped process	*	*	*		*	8
B, is uncapped process	*		*	*		6
C, capped process without deep-UV exposure	*	*			*	0
D, uncapped process without deep-UV exposure	*			*		1
E, uncapped process without AZ development		*	*	*		0
F, control wafers, Al dots deposited on fushly grown SiO$_2$						1

From the results shown in Table XI, the basic novolak/PMMA system will show a low pinhole density provided that the top imaging layer is pinhole free. This also applies to 3-layer systems. The pinhole density in the planarizing layer can be neglected. Whereas there is a possiblity of some pinholes in the middle isolation layer, the possibility of completely transferring these pinholes through the entire planarizing layer is low. In the case of RIE PCM, unless the pinholes in the imaging and the isolation layers line up, the resist in the imaging layer serves as a partial etch mask during O_2 RIE so that if the etching of the planarizing layer is not significantly over end point, the pinhole will be only partially etched. This may not have an adverse effect on the RIE of the substrate if the etch rate ratio is favorable. For lift-off, wet etch, or plating, the partially opened pinholes are not reproduced in the finished pattern. In the case of deep-UV PCM, pinholes in the isolation layer are not reproduced at all, if the imaging layer is a good deep-UV mask. They will be partially reproduced into the planarizing layer, because the imaging layer attenuates the blanket exposure. Therefore the pinhole density of three-layer systems is slightly higher than that of the two-layer organic deep-UV PCM system, depending on the quality of the isolation layer. To date, only the pinhole density of the spun-on-glass material has been published (*39*). It was reported to be less than 3 per cm^2 and therefore does not seem to be significant considering the improbability of their transfer to the substrate.

However, the pinhole density in the imaging layer has to be reduced. This can be done by searching among all the commercial novolak - based resists for an acceptable candidate or by setting an MLR specification for resist vendors improve their quality control. Because thin resists have been used for mask making with an acceptable defect level, no fundamental pinhole problem is anticipated.

6.4.c Additive Defects. In addition to pinholes which are subtractive defects, additive defects in the resist system can also be problematic. Spun on materials usually can be filtered through submicrometer filters to remove particulates. Therefore, with a reasonably clean work station, additive defects can be controlled. However, when another deposition technique is practiced, e.g., evaporation or sputtering of the inorganic resist layer or plasma CVD of the isolation layer, sizable chunks of material can be formed by agglomeration during flight from the source to the wafer. Film deposited on the inside walls of the deposition apparatus can flake off. These types of defects are more difficult to prevent than those related to the spin coating technique. The problem caused by these additive defects is mostly in the areas intended to be opened where they function as an opaque spot to image transfer as seen in Figure 31. Figures 31b and 31d show that with deep-UV delineation of the planarizing layer, the defect will be reproduced unless the size is smaller than the diffraction limit (Figure 31c). This results from the fact that the slope of the delineated images in the planarizing layer is different for the small and large defects because of diffraction.

When two-layer Ge-Se inorganic resist systems are used, further complications can take place as seen in Figure 32. When no shadowing is induced by the additive defect, no harm is done. However, when shadowing takes place, the silver in the shadowed area not only is not photodoped but is depleted because of the edge sharpening effect. During development, the defect is removed, leaving an open undoped area for the developer to further remove the normal Ge-Se material to form a pinhole.

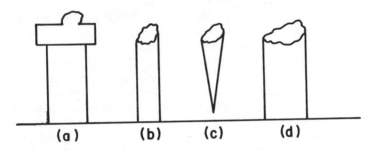

Figure 31.   Effect of additive defects in the isolating layer.   (a) Defects in the unopened areas are acceptable.   (b) Result of RIE in the opened areas. Defects are reproduced.   (c) Diffraction effect eliminating small defects when deep-UV exposure is used.   (d) Diffraction effect reproducing larger defects.

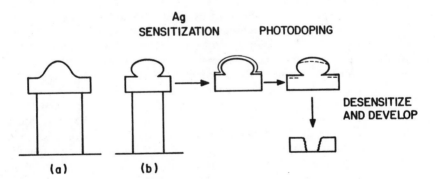

Figure 32.   Effects of additive defects in the unopened areas of a two-layer Ge-Se system.   The situation in (a) is acceptable.   The situation in (b) leads to a pinhole.

6.4.d Interfacial Layers. When one polymer material is spun on another one with a common solvent system, an interfacial layer can be formed and can have a significant impact on the subsequent processing. Among the MLR systems, this interfacial effect is found in the two-layer novolak deep-UV PCM system, the two-layer wet etch PCM system, and the three-layer spun-on-glass RIE PCM system. Fortunately, the solvent for the resist being spun on may dissolve the underlying layer only very slowly, and therefore most of the interfacial mixing occurs during prebaking and not during spinning. The interfacial layer mixing was studied in detail (*26*) for the AZ/PMMA system. The results given as follows can also help to understand interfacial layer mixing of other systems.

A simple test to estimate the interfacial layer thickness is to measure the thickness of the bottom layer before and after spinning, exposure, and development of the top layer. The difference is taken to be the thickness of the interfacial layer for comparison purposes. In reality, the mixing is continuous and the development of the top layer stops inside the interfacial layer instead of at its edges precisely. Furthermore, the test in Reference 26 relies on the IBM Film Thickness Analyzer to measure the resist thickness for convenience. Since this tool operates on the principle of spectral reflectivity changes caused by film thickness changes, a uniform refractive index is important. When some part of the interfacial layer still remains, the measurement can be erroneous in principle.

The AZ/PMMA intermixing was found to be similar to a diffusion process. The thickness of the interfacial layer increases as a function of prebake temperature as shown in Figure 33 where the resist concentration and spin speed are kept constant at a 60% AZ1350J dilution and 8000 rpm respectively. The interfacial thickness also increases with resist concentration at constant spin speed and prebake temperature as seen in Figure 34. Even though the main cause of intermixing is thermal diffusion, some intermixing during spinning was observed as seen in Figure 35 where the prebake temperature and resist concentration were kept constant at 85°C and 60% respectively but the spinning speed was varied. The thickness of the interfacial layer is seen to be a decreasing function of spin speed. Another observation, not plotted here, is the reduction of the interfacial layer thickness by dynamic spinning, i.e., the resist is applied after the wafer has attained the desired spin speed instead of creating a static resist puddle on the wafer before starting the spinner. After choosing a prebake temperature designed to satisfy the dissolution-rate contrast requirement of the resist, the resist concentration and spin speed can be optimized for a minimum interfacial layer thickness of 12 nm as shown in Figure 34, where the optimum point of a 60% concentration at 5400 rpm was chosen to produce a 0.4 - μm AZ thickness with a 85°C prebake temperature.

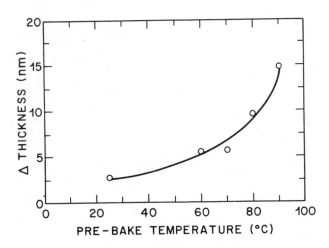

Figure 33.   The interfacial layer thickness as a function of AZ1350J pre-bake temperature with an AZ dilution of 60% and an 8000-rpm spin speed, resulting in a 0.32-μm film.

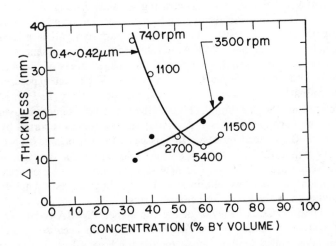

Figure 34.   The circles show interfacial thickness as a function of resist concentration spun at 3500 rpm and prebaked at 85°C.   The solid dots show interfacial thickness also as a function of concentration but the spin speed is changed to maintain a constant resist thickness of 0.41 ± 0.01 μm.

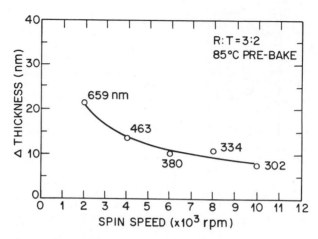

Figure 35.   The interfacial layer thickness as a function of spin speed.

Even after minimizing the interfacial layer, special processing steps have to be used to eliminate its effects. For the uncapped process, the interfacial layer causes incomplete removal of the undeveloped AZ film in parts of the wafer whereas for the capped process, the interfacial layer prevents the PMMA developer from penetrating as seen in Figure 36 where only pinholes in the interfacial layer allowed the chlorobenzene developer to develop the PMMA resist. The facilitate cap removal, a presoak in 1:1 methanol: water before PMMA development is recommended. For the capped process, the interfacial layer in the opened areas of the AZ resist, is removed with an O_2 plasma before the deep-UV blanket exposure. Santini and Viswanathan (*46*) used a $CF_4 + O_2$ plasma to treat the PMMA surface before coating the AZ film. This resulted in a very effective layer separation for the uncapped process. A similar technique was reported by Chao et al (*55*) for the capped process. In this case, the O_2 plasma step before the deep-UV exposure listed as step 6b in Table III can be omitted. However, the post-exposure bake of step 5a now serves to improve AZ adhesion during development as well as elimination of standing waves. Therefore, it is required even in the case of e-beam imaging.

The interfacial behavior of the copolymer/PMMA two-layer wet etch PCM system is similar to the AZ/PMMA system because of an identical bottom layer and a similar solvent for the top layer. In this case, consideration of the interfacial layer leads one to avoid using completely mutually exclusive developers. The developer for the bottom planarizing layer should be chosen such that it also develops the top layer at an low dissolution rate so that the interfacial layer can be broken through during the second

Figure 36. SEM showing the interfacial layer preventing the uncapped developer from removing PMMA. Only pinholes in the interfacial layer could be developed.

development. The penalty paid for this choice is that the linewidth in the imaging layer continues to change during the development of the planarizing layer and has to be anticipated in advance.

Interfacial mixing (*17,18,39,40*) can take place between the imaging and the isolating layers or between the isolation and the planarizing layers. Interfacial mixing can be extremely disastrous when a novolak - based resist is used as the planarizing layer, because the solvent of the spun-on-glass is mainly ethyl alcohol. Therefore, this planarizing layer has to be hard baked above 200°C to ensure complete cross-linking and insolubility of the novolak resist. In order to minimize interfacial mixing between the imaging and isolating layers, a proper baking temperature again has to be used for the spun-on-glass layer. The temperature selection is now more critical, because low temperature can result in interfacial mixing, whereas high temperature can cause the isolating layer to crack.

Figure 37.   Residue on the wafer surface after O_2 RIE.

6.4.e Etch Residue. Generally, dry etching produces more etch residue than wet etching. Therefore, when RIE is used for the middle or bottom layer, great care should be taken to prevent residue formation or to remove the residue after RIE. The residue is usually caused by sputtering of the electrode material and can be prevented by covering all exposed conducting surfaces with a dielectric such as quartz or even photoresist. The chamber should be maintained clean by properly regulating the type of etched materials in each RIE tool. It is also not advisable to mix resist etching and substrate etching in the same tool. Once residue is formed, it is difficult to remove by any other dry etching technique but is quite simple to clean up with rinsing or a mild wet etch followed with a rinse. Figure 37 shows the residue formed after the bottom resist layer was etched by O_2 RIE. The middle layer used was Si.

6.4.f Film Stress. Because many layers of different materials are involved in an MLR system, stress is a major concern. Stress problems occur most frequently in three-layer systems because of the isolation layer. Figure 38 shows randomly occurring deformation of the Si film resulting from incorrect deposition conditions. Sometimes, the film can be perfect after deposition but will crack after application of the imaging layer. Ti is quite susceptible in this regard. For the spun-on-glass system, cracking of the glass layer takes place when the baking temperature is too high. On the other hand, a low baking temperature can cause interfacial mixing problems between the imaging and the glass layers as discussed in Section 6.3.c. Even with a two-layer organic system such as the novolak deep-UV PCM system, stress can cause a problem. Figure 39 shows a bent AZ/PMMA line caused by a 160°C bake prior to PMMA development. The intent was to completely cross-link the AZ layer so that it could not be removed by MIBK, the uncapped developer, to produce a capped image. The 160°C bake obviously induced too much stress and made this process unusable.

6.4.g Interference Effects. As discussed in Section 6.3.a, multiple optical interference effects from wafer topography can be eliminated with a highly absorptive planarizing layer. However, optical interference can still take place above the planarizing layer. In a 2-layer system, the top surface of the imaging layer and the surface between the imaging and the planarizing layers can produce reflections. In a 3-layer system, 3 surfaces are effective in contributing to optical interference effects. Unless the refractive index and the absorption coefficient of these layers are identical, multiple reflections are inevitable. Even though it is possible to choose materials of similar refractive indices, it is not possible to match the absorption coefficients, because the absorption in the planarizing layer has to be higher to facilitate isolation of wafer topography. Therefore, if this is left untreated, the multiple reflections form standing waves in the imaging layer. In the case of a highly reflective isolation layer such as Al, the standing wave can be very severe just as in the case of SLR. A prevention technique for this interference phenomenon is not known. However, a post-exposure bake of the imaging layer can smooth out the sharp variation in dissolution rate in the exposed and unexposed vertical dimension with a slight penalty of contrast loss laterally. A developer that has a high dissolution rate for the unexposed resist can be used to break through the nodes of the standing waves. For example, an undiluted AZ developer can be used instead of the recommended 1:1 dilution form. Of course, the *resist development* contrast suffers.

6.4.h Spectral Transmission for Exposure and Alignment. Spectral transmission is of concern in two situations. In a deep-UV PCM system, the opacity of the exposure PCM should be sufficiently high to facilitate a high masking contrast. However, it is desirable that this masking material

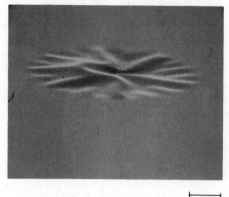

5μm

5μm

Figure 38. Blister formation in a hydrogenated amorphous Si Film deposited by plasma CVD.

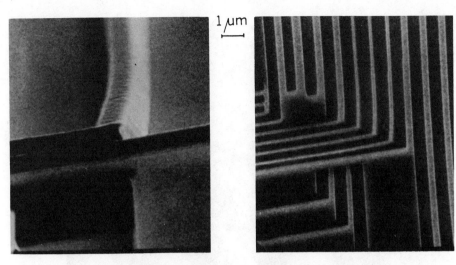

Figure 39.   A bent isolated line resulting from a 160°C 1-hour bake of the AZ image on PMMA. Reducing either the baking time or temperature resolved the problem.

show sufficient transmittance at the alignment wavelength for viewing of the alignment mark. The deteriorated image in the planarizing layer caused by an insufficiently opaque deep-UV PCM is shown in Figure 40. In the case of RIE PCM systems, masking opacity is obviously not important, but transparency for alignment is desired. SiO_2 is a good material for the RIE PCM systems because of its transparency. However, when the planarizing layer is dyed, the dye should again be sufficiently transparent at the alignment wavelength. Si and novolak resists are good deep-UV PCM materials because of their opacity in the deep-UV and transparency at *longer* wavelengths. Ge-Se is unique in that it is opaque at the imaging wavelengths yet transparent in longer wavelengths. Normally, if a resist is too absorptive at the imaging wavelengths, *light* cannot penetrate through its entire thickness for a complete delineation to the bottom. *However*, for Ge-Se, only the very top surface has to be photodoped. Because of a phase separated columnar structure of the Ge-Se layer, anisotropic etching is observed when the Ge-Se film is subjected to a bicomponent etchant consisting of a mildly concentrated base solution, and a diluted Na_2S solution (*56*). Figure 41 shows the vertical sidewalls of a unsensitized Ge-Se film etched using PMMA as the wet etch mask for the bicomponent etch solution.

Figure 40. Uncapped resist image showing insufficient deep-UV absorption in the PCM.

When the resist system is opaque at the alignment wavelength, adjusting the spectral distribution cannot help. Even when the resist system is not opaque, multiple reflection by the MLR structure can cause a loss in the alignment signal and create ghost alignment images from the mirror images of the alignment mark created by each reflection. This is also true for SLR systems because of the substrate films on the wafers. While the substrate film stack is not controlled by the lithographer, the resist structure can be removed entirely if necessary. One should bear in mind that though removing the entire resist structure helps most of the time, it may be undesirable in special situations. For example, in the simple case of resist on SiO_2 on Si, if the SiO_2 thickness is such that annihiliation of reflection takes places without the resist layer, removing the resist layer simply reduces the alignment signal.

A straightforward scheme to remove the resist on the alignment marks is to go through an alignment-exposure-development procedure using

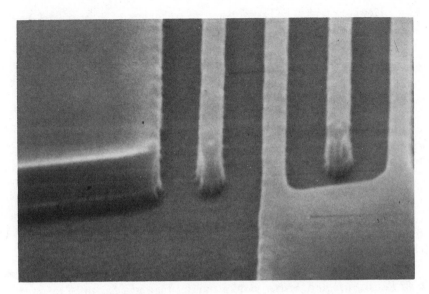

*Figure 41.   Unsensitized and unexposed Ge-Se image that was anisotropi-
cally wet etched using PMMA as the wet-etch PCM.*

a special mask. This works satisfactorily for single layer resist and the
copolymer wet etch PCM system. For the novolak deep-UV PCM, the bottom layer has to be delineated at the alignment site before application of the
imaging layer. This inevitably has an adverse effect on pinhole density.
When an inorganic layer is present, or when the imaging layer is a negative
resist,   removing   this   structure   at   the   alignment   site   by
exposure/development is not possible.

6.4.i Inspection.   Because of the small thickness and transparency requirement for alignment, the delineated imaging layer is difficult to inspect using
visible light, prior to delineating the image in the bottom layer. The contrast of the image is low in visible light. Optoelectronic linewidth measurement becomes even more difficult because of the higher image contrast
required.   Scanning electron microscopic inspection is also difficult because
of the shallow relief image in the top layer. The thick underlying polymeric
film can often be distorted by the e-beam. Bubbling of the bottom layer can
take place in severe cases.

However, typical end point detection techniques (57) for the development of the imaging layer can still be used. In particular, monitoring of multiple reflections due to film thickness change, as shown in Figure 42, is applicable. The fringe count is only between 1 and 2; therefore, the determination of end point and the percentage over development beyond end point have to be assessed quickly. We have found the double exposure end point detection (DEEPDET) (58) technique very convenient to use with MLR systems. The principle of DEEPDET is illustrated by Figure 43. A small end point site on the wafer is first blanket exposed, then pattern exposure follows, or vice versa. In the doubly exposed end point site, the part of the resist that receives more exposure is quickly cleared as shown in Figure 43c. The remaining part is developed at the same rate as the exposed resist in the pattern area. The complete clearing of the resist in the end point site indicates the development end point as shown in Figure 43c. Development continues to a predetermined percentage over the end point. Alternately, the blanket exposure can be reduced to a precalibrated dosage so that the clearing of the resist at the end point site signifies completion of development.

6.4.j Resist Stripping.   After pattern transfer the resist has to be removed. For the all organic MLR systems, the resist layers can be removed with an O_2 plasma if the top surface is not severely altered after an ion implantation or a substrate etching. Otherwise, a solvent treatment followed with clean-

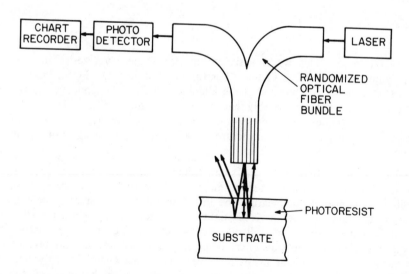

Figure 42.   Schematic drawing of a laser resist-end-point detector.

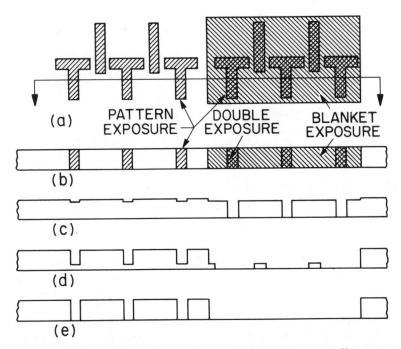

Figure 43.   The principle of operation of DEEPDET. A small area on the wafer is first exposed, then a pattern exposure follows. The complete clearing of resist in this doubly exposed test area signifies the development end point.

ing by a sulfuric-nitric acid mixture can be used. After metal lift-off, the resist is removed by the solvent used in the lift-off process, and no further stripping is necessary. Any acid exposure is undesirable for metal. For the MLR system containing an inorganic layer above the planarizing layer, O_2 plasma can be used only if the inorganic layer was first removed. In most cases, the inorganic layer is removed during substrate etching. Otherwise an undesirable additional step has to be used to remove the inorganic layer. alternately, solvents can be used to remove the resist without requiring removal of the inorganic layer, but this applies only to the cases where the planarizing layer was not cross-linked to become insoluble. Sulfuric-nitric acid stripping seems to be the only practical solution for a cross-linked planarizing layer, but cannot be used to lift-off metal. Then a fourth soluble layer is required under the planarizing layer to facilitate lift-off.

Generally, all organic MLR systems are more favorable when resist stripping is concerned but cannot sustain the high processing temperatures of inorganic containing MLR systems whose planarizing layer can be cross-linked for temperature stability which will be further discussed in Section 6.5.b.

6.5 Comparison of Existing Systems

To clarify the selection of a particular MLR system (1LR, 2LR, or 3LR system) a comparison in terms of process complexity, resolution, aspect ratio, linewidth tolerance, sensitivity and effort required for research and development will be given. Then a comparison between deep-UV and RIE PCM systems in terms of resolution, aspect ratio, substrate patterning processes allowed, temperature stability, resist removal at the alignment sites, tool-controlling parameters, and tool cost will be included.

6.5.a Comparison of 1LR, 2LR, and 3LR Systems.

6.5.a.1 Process Complexity. It is obvious that one of the attractive aspects of 1LR systems is process simplicity. Even though the improvement of 1LR systems requires additional processing steps as discussed in Section 6.3.d, it is still simpler to process 1LR systems. Four processing steps are required from spin coating to development of a basic 1LR system. Two additional steps are required to process the soaked system. For two-layer systems, the wet etch PCM and the siloxane RIE PCM systems are the simplest requiring a total of seven processing steps. However, reactive ion etching of the planarizing layer for the siloxane system should be considered more complicated than the wet development step in the wet etch system. The most elaborate capped deep-UV PCM systems require eleven steps. The two-layer Ge-Se system requires only eight steps. It should be noted that certain processing steps involved in the Ge-Se system are much more complicated than others, e.g. the Ge-Se deposition and RIE steps. The two basic three-layer systems require eleven and twelve steps for RIE PCM and deep-UV PCM respectively. The former can have one or two complicated steps depending on whether spun-on-glass is used. Similarly, the latter has zero or one complicated step. Obviously, the image exposure step is also complicated but has not been counted here because it is universally required for any resist system.

6.5.a.2 Resolution and Aspect Ratio. With all the reasonings given in Section 6.2, it is now obvious that with 1LR, either the aspect ratio is unacceptable for a given resolution or the resolution is too low for a required aspect ratio. The merit of 2LR and 3LR systems is the ability to improve either or both of their aspects using an identical imaging tool. Further differentiation of the resolution and aspect ratio in MLR systems depends on the type of PCM used and will be discussed later.

6.5.a.3 Linewidth Tolerance. The linewidth variation of the final fabricated feature on the wafer consists of contributions from the dimension variation on the mask patterns, variation of the resist image, and variation of the etched or lift-off image. As discussed in Sections 6.2.a and 6.3.d, when a lift-off or an anisotropic etching process is used, the resultant fabricated feature on the wafer follows the dimension of the top of the resist image. Therefore, in a two-layer structure, as long as the opening in the imaging layer is smaller than that in the planarizing layer and is a sufficient lift-off or anisotropic etch mask, the dimension variation in the planarizing layer has no first order effect on linewidth tolerance. However, with a three-layer system, the critical dimension layer is often transferred from the imaging layer to the isolation layer. Therefore, an additional processing variation is incurred.

6.5.a.4 Sensitivity Potential.   The choice of materials for the 1LR system is limited because it is difficult to satisfy the sensitivity, resolution, and aspect ratio requirements plus all other processing characteristics such as thermal stability, absence of pinholes, stripability, etc. Improvements in sensitivity will depend on exploration of new materials. The two-layer system relies on special combination of materials. Therefore, the sensitivity depends on the combination chosen. For example, the siloxane system has a high sensitivity, whereas the Ge-Se system still needs improvement in sensitivity. The novolak deep-UV PCM system has a sensitivity comparable to conventional 1LR systems.

6.5.a.5 Efforts on Research and Development.   Development of 1LR systems depends on the exploration of new materials, and therefore is the most challenging. This can be substantiated by the volume of literature dedicated to 1LR resist material research, yet only a few successful materials have been actually used. Though there is no guaranteed success, once an acceptable material is developed, it would readily find application in manufacturing. However, even if such a superior material is developed, it can at most be used on planar substrates because of the optical interference effects from the wafer topography or the e-beam proximity effects. The x-ray oblique profile phenomenon and the short ion beam stopping distance also demand MLR systems despite probable future sensitivity achievement in 1LR systems. Research and development of 2LR systems are heavily process oriented once a promising set of candidates is identified. Material replacement in either layer leads to new elaborate process refinements. However, once a reliable process sequence is defined for a correctly chosen pair of candidates, process simplicity and a tighter linewidth tolerance would make the 2LR system more attractive than 3LR systems. The research and development of 3LR systems is mainly focussed on identification of acceptable materials for the isolation layer and the RIE process for the isolation and the planarizing layer. Once this is achieved, a large variety of resist materials can be used for the imaging layer. Therefore this is a relatively easier system to develop and is more universal than 2LR systems. However, manufacturing cost due to process complexity and a larger linewidth tolerance can negate the savings in development cost.

The comparison is outlined in Table XII. One - layer systems clearly are favored for manufacturing if the performance in sensitivity, resolution, aspect ratio, and linewidth control can be improved. A high research effort is required and can result in success mostly for low topography applications. Two - layer systems perform better than 1LR systems because of planarization, antireflection, reduction of the imaging resist thickness, and even - edge sharpening for special systems. However, they are limited to specific resist combinations therefore, renewed intense research efforts have to be devoted to each new entry. They will be favored by manufacturers over 3LR systems because of process simplicity and linewidth control. Three - layer systems suffer from process complexity and slightly worse linewidth control than 2LR systems. However, after the research is done on the isolation and the planarizing layers, they are easily adaptable to any new imaging resist; therefore research of 3LR systems efforts become less demanding. The chance for manufacturing application lies in the sensitivity potential which may increase aligner throughput to compensate for only disadvantage in process complexity.

TABLE XII. Comparison of 1LR, 2LR and 3LR Systems

| | Process Complexity | | |
	4-6 Steps	6-11 Steps	11-12 Steps
Resolution & Aspect Ratio	Limited by Resist Performance	Resist Performance Enhanced	Resist Performance Enhanced
Linewidth Tolerance	Keyed to 1 Layer	Keyed to 1 Layer	Often keyed to 2 Layers
Sensitivity Potential	Limited, Depending on Resolution & Aspect Ratio Requirements	Less Limited than 1LRs, Depending on Special Resist Combinations	Most Unlimited, Only Limited by AvailABLE Resist Sensitivity
Efforts on Research & Development	High, Material Oriented, Unlikely to Succeed	High, Processing Oriented, Some Successful Combinations have been Reported	Not as High, Provides a Universal System. Advantage Likely to be negated by Manufacturing Cost

6.5.b Comparison of Deep-UV and RIE PCM System.

6.5.b.1 Resolution and Aspect Ratio. Deep-UV exposure through the PCM is just like proximity printing with a mask intimately contacted to the resist of the planarizing layer. Optical diffraction and attenuation cause the patterned, exposing raidation to be dispersed and absorbed respectively beyond a certain distance from the mask, with the result that the delineated image can no longer be maintained within the specified tolerance limit. With PMMA films, an aspect ratio of 4 can be achieved for 0.5 - μm features (*46*). For feature sizes smaller than 0.5 μm, a lower aspect ratio has to be used. For larger features, the aspect ratio improves, but the limit has not been established experimentally. The theoretical work can be found in Section 215 of Reference 56. The aspect ratio limit for the RIE PCM system is not as well understood. Perhaps the reason for the lack of study is absence of necessity. The aspect ratio of RIE images has consistently been shown to exceed lithographic requirements. For example, submicrometer features with extremely high aspect ratios can be found in Figures 15 and 16. The resolution limit of an RIE system is not known either. In the range of features sizes 0.2 μm and larger, a properly optimized RIE process can produce a seemingly endless height of the etched resist image in the planarizing layer.

6.5.b.2 Feature-Patterning Process Allowed. PMMA, PMMA copolymers, and polymethyl isopropenyl ketone (PMIPK), have been used for the planarizing layer of the deep-UV PCM systems. To date, PMMA has been most widely used. Therefore the discussions on the substrate-patterning process will be based on PMMA.

Both the deep-UV and the RIE PCM systems can be used for lift-off as exemplified by the work in the Lift-Off era using RIE PCM's and the results shown in Ref. 26. In this case, the deep-UV PCM system is more desirable because the resist can be removed readily by using regular solvents. The planarizing layer of the RIE PCM system is more difficult to remove because of cross-linking produced by the hard baking. Very often a fourth layer is used to facilitate resist removal. In addition, the RIE process has to include an isotropic component to produce an overhang in the RIE masking layer.

Both the deep-UV-and the RIE PCM-delineated resist structures can be used for electroplating which was successfully demonstrated by Satini and Viswanathan (*45*) as shown in Figure 22 for the deep-UV PCM case. In this case the control of the image in the planarizing layer is critical. Because of the unique anisotropic characteristics of RIE, this control should also be feasible for RIE PCM. In such cases linewidth variations include the contribution from the planarizing layer as well as top layers. The consideration of resist removal after plating is similar to the case for lift-off.

In substractive etching applications where the resist image is used as an RIE mask for the underlying features, resist etch rate ratio with respect to that of the underlying features is crucial. The PMMA resist is known to have a higher etch rate than novolak based resist, therefore it may not be able to provide a sufficient etch barrier. It has been shown that using the novolak cap as the RIE mask in a capped deep-UV PCM system, it is feasible to RIE Si_3N_4, SiO_2, and poly Si films for MOSFET applications (26). The PMMA thickness in uncapped deep UV PCM system is also sufficient for this application. However, Al etching with the deep-UV PCM system has not been feasible.

6.5.b.3 Temperature Stability. The planarizing layer in RIE PCM systems is often baked to induce complete cross-linking to facilitate coating of other layers. This also enables the final resist structure to withstand a processing temperature up to 220°C without deformation of the resist image profile. On the other hand, the deep-UV sensitive planarizing layer in the deep-uv PCM systems has to remain sensitive until after delineation of the planarizing layer and cannot readily be cross-linked. Even though PMMA and its copolymers have to be prebaked at temperatures up to 200°C to promote contrast, they have much lower glass transition temperatures. To maintain the resist image profile with processing temperatures above 160°C would be difficult. PMMA without copolymeration exhibits even a lower temperature tolerance.

6.5.b.4 Alignment Site Removal. The situations that call for removal of the alignment sites have been discussed in Section 6.4.h. When this is necessary, an all - organic system is desirable. Any system that requires an RIE step is virtually impossible to clear unless some elaborate steps are taken.

6.5.b.5 Tool Controlling Parameters and cost. To delineate the planarizing layer, the deep-UV scheme requires control of the exposure spectrum, exposure dosage, and the development end point. With the RIE scheme, the geometry and cleanliness of the etch chamber and electrodes, the composition of the etchant gas, pressure, flow rate substrate temperature, DC bias, RF power level and elimination of the reflected RF power have to be considered. Therefore the RIE scheme is more complicated and has a higher probability of failure. Also, RIE tools are more costly to purchase, operate, and maintain. They also require vacuum systems and therefore more floor space. Conversely, a deep-UV blanket exposure tool costs about 1/10 of the RIE tool, is compact, and is easy to maintain and operate.

The comparison is outlined in Table XIII. RIE PCM is preferred for Al etching and for situations in which high temperature is required. Diffraction limits do not cause a problem for anticipated submicrometer fabrication. Deep-UV exposure tools are much less expensive to purchase, maintain, and operate. There are also much fewer control parameters for the deep-UV scheme.

TABLE XIII. Comparison of Deep-UV and RIE PCM Systems

Parameter	Deep-UV PCM	RIE PCM
Resolution & Aspect Ratio	Diffraction limited, 0.5 μm features on 2 μm bottom layer	Anisotropy limited, Limits beyond lithographic requirements not yet established.
Substrate-Patterning Process Allowed	Lift-off, Plating, RIE of Si_3N_4, SiO_2 and Si	Lift-off, (conditional) Plating, RIE of Si_3N_4, SiO_2 and Si,Al
Temperature Stability	160°C	220°C
Resist Removal for Alignment	Feasible	Difficult
Tool Parameters	Exposure spectrum Exposure dosage Development end point	Chamber & electrode Cleanliness, Gas composition, Pressure, Flow rate, Substrate temperature, DC bias, RF power, RF reflection,
Tool Purchase Maintenance & Operation	Negligible	Costly

6.6 Conclusion

The various types of multi-layer resist systems differ significantly in terms of their advantages, the choice of existing systems, and practical considerations. Among the existing systems, the three-layer systems have found many applications because of their versatility. The two-layer novolak deep-UV PCM systems have also been used widely because of their inherent adaptivity to conventional processing. The two-layer Ge-Se systems offer the possibility of fabricating a 0.2 - μm device with existing optical mask aligners. The two-layer siloxane systems have shown high e-beam and deep-UV sensitivity. Further work must be pursued to overcome the shortcoming of each system and to take full advantage of the MLR performance.

ACKNOWLEDGEMENT

The author is indebted to the authors and their sponsoring corporations who granted permission to reproduce graphics and figures of their work. Carlton Osburn generously reviewed this chapter in detail. The excellent typing skill of Patricia Ross is also extremely appreciated.

Literature Cited

1. Lacombat M.; Dubroeucq G. M. *Proc. SPIE*, 1979, *174*, p 28.
2. Walker, E. J., *IEEE Trans. Electron Devices*, 1975, *ED-22*, p 464.
3. Kyser, D. S.; Viswanathan, N. S.; *J. Vac. Sci. Technol.*, 1975, *12*, p 1305.
4. Greeneich, J. S.; Van Duzer, T. J.; *Vac. Sci. Technol.*, 1973, *10*, p 1056.
5. Everhart, T. E., *J. Appl. Phys.*, 1960, *31*, p 1483.
6. Greeneich, J. S., *Electrochem. Soc.*, Extended Abstract, 1980, *80* (1), p 261.
7. Thompson, L. F., *Solid State Technol.*, 1975, *17* (8), p 41.
8. Murase, K.; Kakuchi, M.; Sugawara, S., International Conference on Microlithography, 1977, Paris, p. 216.
9. Somar Manufacturing Co., Ltd.
10. Kakuchi, M.; Sugawara, S.; Murase, K.; Matsuyama, K. *J. Electrochem. Soc.*, 1977, *124*, p 1648.
11. Moran, J. M.; Taylor, G. N., *J. Vac. Sci. Technol.*, 1979, *16*, p 2014.
12. Taylor, G. N.; Wolf, T. M., *J. Electrochem. Soc.*, 1980, *127*, p 2665.
13. Ryssel, H.; Haberger, K.; Kranz, H., *J. Vac. Sci. Technol.*, 1981, *19*, p 1358.
14. Franco, J. R.; Havas, J. R.; Levine, H. A., US Patent 3,873,361, 1973.

15. Grebe, K.; Ames, I.; Ginzberg, A. *J. Vac. Sci. Technol.*, 1974, *11*, p 458.
16. Chamberlin, P. D.; Bergeron, R. J.; IBM technical memos, 1975.
17. Franco, J. R.; Havas, J. R.; Rompala, L. J., US Patent 4,004,044, 1975.
18. Havas, J. R. *Electrochem. Soc.*, Extended Abstracts, 1976, *76* (2), p 743.
19. a. Hatzakis, M. US Patent 4,024,293, 1977.
    b. Hatzakis, M. *J. Vac. Sci. Technol.*, 1979, *26*, p 2984.
    c. Grobman, R. D.; Luhn, H. E.; Donohue, T. P.; Speth, A. J.; Wilson, A.; Hatzakis, M. *IEEE Trans. Electron Devices*, 1979, *ED-26*, p 360.
20. Carruthers, R.; Nagasaki, T.; Poponiak, M. R.; Zielinski, L. B. *IBM Technical Disclosure Bulletin*, 1976, *19*, p 1214.
21. Logan, J. S.; Schwartz, G. C.; Zielinski, L. B. *IBM Technical Disclosure Bulletin*, 1977, *20*, p 1024.
22. Logan, J. S.; Mauer, J. L.; Zielinski, L. B. *IBM Technical Disclosure Bulletin*, 1979, *21*, p 4865.
23. a. Lin, B. J. SPSE 31st Ann. Conf. Washington, D.C. 1978.
    b. Lin, B. J. *J. Electrochem. Soc.*, 1980, *127*, p 202.
24. Lin, B. J. Proceedings of SPIE, 1979, *174*, p 114.
25. Lin, B. J.; Chang, T. H. P. *J. Vac. Sci. Technol.*, 1979, *16*, p 1669.
26. Lin, B. J.; Bassous, E.; Chao, V. W.; Petrillo, K. E. *J. Vac. Sci. Technol.*, 1981, *19*, p 1313.
27. Tai, K. L.; Johnson, L. F.; Murphy, D. W.; Chung, M. S. *C. Electrochem Soc.*, Extended Abstract 1979, *79* (1), p 244.
28. Tai, K. L.; Sinclair, W. R.; Vadimsky, R. G.; Moran, J. M.; Rand, M. J. *J. Vac. Sci. Technol.*, 1979, *16*, p 1977.
29. Yoshikawa, A.; Ochi, O.; Nagai, H.; Mizushima, Y. *Apply. Phys. Lett.*, 1976, *24*, p 677.
30. Vadimsky, R. G.; Tai, K. L. *Electrochem Soc.*, Extended Abstract 1980, *80* (2), p 824.
31. Hatzakis, M.; Paraszczak, J.; Shaw, J. *Proceedings of the Internal Conference on Microlithography*, 1981, p 386.
32. Roberts, E. D. *Phillips Techn. Review*, 1975, *35*, p 41.
33. Gazard, M.; Dubois, J. C.; Duchesne C. *Appl. Polym. Sympos.*, 1974, *23*, p 107.
34. Moran, J. M.; Maydan, D. *J. Vac. Sci. Technol.*, 1979, *16*, p 1620.
35. Watts, R. K.; Fichtner, W.; Fuls, E. N.; Thibault, L. R.; Johnston, R. L. *Proceedings IEDM*, 1980, p 772.
36. Kruger, J. B.; Rissman, R.; Chang, M. S. *J. Vac. Sci. Technol.*, 1981, *19*, p 1320.
37. Tennant, D. M.; Jackel, L. D.; Howard, R. E., Hu, E. L.; Brabbe, P., Capik, R. J. *J. Vac. Sci. Technol.*, 1981, *19*, p 1304.

38. O'Toole, M. M.; Liu, E. D.; Chang, M. S. *Proceedings of SPIE*, 1981, *275*, p 128.

39. Ray, G. W.; Peng, S.; Burriesci, D.; O'Toole, M. M.; Liu, E. D. *J. Electrochem. Soc.,* 1982, *129*, p 2152.

40. Ting, C. H.; Avigal, I.; Lu, B. C. *Proceedings of Kodak Microelectronics Seminar,* 1982.

41. Ong, E.; Tai, K.; Reichmanis, E.; Wilkins, C. W. Jr., *Proceedings of Kodak Microelectronics Seminar,* 1981, p 91.

42. Vadimsky, R. G.; Tai, K. L.; Ong, E. *Proceedings of the Symposium on Inorganic Resist Systems,* 1982, *82* (9), p 37.

43. Batchelder, W. T.; Takemoto, C. *Semiconductor International,* 1981, *7*, p 213.

44. Griffing, B. F.; Johnson, P. D.; Norton, J. F. *Proceedings of SPIE*, 1981, *275*, p 136.

45. Ury, M. G.; Matthews, J. C.; Wood, C. H. *Proceedings of SPIE*, 1982, *334*, p 241.

46. Santini, H.; Viswanathan, N. W. *Proceedings of Kodak Microelectronics Seminar,* 1982.

47. Lin, B. J.; Chao, V. W.; Kaufman, F. B.; Kramer, S. R. *Electrochem. Soc.,* Extended Abstracts, 1982, *82* (1), p 275.

48. Bassous, E.; Ephrath, L. M.; Pepper, G.; Mikalsen, D. J. submitted for publication, 1982.

49. Hatzakis, M. IBM *J. Res. Develop.,* 1980, *24*, p 452.

50. Matsuzawa, T.; Tomioka, H. *IEEE Trans. Electron Devices,* 1981, *ED-28*, p 1284.

51. Chao, V. W.; Fan, B.; Lin, B. J. recent results.

52. Rothman, L. B. *J. Electrochem. Soc.,* 1980, *127*, p 2216.

53. Bassous, E.; Pepper, G. *IBM Research Report,* RC 9480, 1982.

54. Fatula, J. private communications, 1980.

55. Chao, V. W.; Pepper, G.; Lin, B. J. recent results. Tai, K. L.; Ong, E.; Vadimsky, R. G.; Kemmerer, C. T.; Bridenbaugh, P. M. *Proceeding of the Symposium on Inorganic Resist Systems,* 1982, *82* (9), p 49.

56. Lin, B. J. "Fine Line Lithography," Newman, R., Ed., North-Holland Publishing Co., 1980, p 209.

57. Lin, B. J. *IEEE Tans. Electron Devices,* 1978, *ED-25*, p 419.

RECEIVED February 14, 1983

INDEX

INDEX

Production by Robin Giroux and Frances Reed
Elements typeset by Service Composition Co., Baltimore, MD
Printed and bound by Maple Press Co., York, PA